ENVIRONMENTAL ECONOMICS AND MANAGEMENT: POLLUTION AND NATURAL RESOURCES

Environmental Economics and Management: Pollution and Natural Resources

Finn R Førsund and Steinar Strøm

CROOM HELM
London • New York • Sydney

Published in the USA by
Croom Helm
in association with Methuen, Inc.
29 West 35th Street
New York, NY 10001

British Library Cataloguing in Publication Data

Førsund, Finn R.
    Environmental economics and management:
    pollution and natural resources.
    1. pollution – Economic aspects
    I. Title   II. Strøm, Steiner
    363. 7'36   HC79. P55
    ISBN 0-85664-900-7

Library of Congress Cataloging-in-Publication Data

Førsund, Finn R.
    Environmental economics and management.

    Includes indexes.
        1. Environmental policy. 2. Pollution — Environmental
aspects. 3. Natural resources. 4. Human ecology.
I. Strøm, Steinar. II. Title.
HC79.E5F64 1987   363.7'0525   87-9179
ISBN 0-85664-900-7

Printed and bound in Great Britain by Mackays of Chatham Ltd, Kent

Contents

Contents

Part I

Introduction

1

Introduction

The main themes of this book are pollution and other external effects, and natural resources; its main aim is to establish the subjects mentioned within the standard setting of economic analysis. We have not tried to write a book surveying the state of the art in this field by collecting the flowers scattered about in the various journals. Instead we have tried to write a book which gives our interpretation of the issues involved.

The general pedagogical principal is that the main problems introduced in each chapter are first set out in fairly non-technical terms. Formal models are first used in Chapter 6. The technical parts of the chapters should be suitable for more advanced students.

In Chapter 2 a brief historical survey of the seminal contributions in this field is given. However, the economists are not the only academics concerned with the deeper problem of the interaction between Man and Nature. We have therefore found it appropriate to quote from the writings of philosophers.

We owe a great debt to our teacher, Professor Trygve Haavelmo, who inspired us through seminars and his own writings to take up the kind of economic research illustrated in this book.

2

A Brief Historical Survey

At the end of the 1960s and the beginning of the 1970s the 'green wave' swept across most of the countries of Western Europe, as well as Japan and the USA. Eastern Europe, too, was not unaffected, even though we are less familiar with conditions in that part of the world. Labels descriptive of the 'surf-riders' who were swept along on the crest of this wave include such phrases as 'conservation of resources', 'the fight against pollution', 'campaigns for decentralisation of decisions', together with touches of populist proposals for the organisation of society, 'many small firms better than a few big ones', 'small is beautiful', 'the struggle for the preservation of ecological variety by means of self-subsistence', etc.

The contents of this book show that some of this criticism can be justified by economic theory. Without collective action, that is, without the intervention of public authorities, the extent and scope of pollution and other ways in which large tracts of common land can be destroyed would be excessive. In an economy without collective control, resources would inevitably be used up far too rapidly.

The material presented in this book, however, shows that a clear distinction exists between the opinion of economists and the causes championed by the 'surf-riders'. The extreme environmental conserver's ideal of *ecological variety* as the overriding goal of society would at least by economists be regarded as an unduly exacting ideal for it to be able to safeguard the interests of the 'general public'. In our role as economists we would like to strike a blow for economic variety; by this we mean that the ideal should be to organise society in such a way as to achieve a balanced and suitable blend of a bit of everything

— to include unspoilt virgin countryside as well as countryside that has suffered as a result of industrial production.

On yet another point we, as economists, have to express disagreement. In this connection it might be appropriate to present some of the views held by the philosopher Arne Næss, who was one of Norway's better-known ecosophists in the 1970s (ecosophy is ecophilosophy with value postulates, ecophilosophy is the philosophy of ecology):

'Ecosophy is among other things a struggle for qualitative versatility, a struggle for variety and life in common, not you or me, but you and me. It is a struggle for equality in the biosphere; all living organisms, animals, human beings, and plants, should be equal and should be given the same opportunities for self-realisation. All organisms are involved in an eco-system.'

This philosophy clearly contains touches of nature mysticism, inspired by the philosopher Baruch de Spinoza (1632-77). Professor Arne Næss has a Norwegian philosophical predecessor in Petter Wessel Zapffe who, in 1961, wrote on the biosophical perspective. As early as the 1920s he mooted ideas which are reappearing in print in ecological quarters. Zapffe, moreover, was a prolific author: in an article in the periodical *Speideren* (*The Boy Scout*) in 1958 entitled 'Farewell, Norway' he puts these words into the mouth of Tinde-Jørgen (Mountain Joe):

'Hearts should, if possible, not be exposed to organised heavy traffic. Right up to the year 1910 there was some point in "opening up the mountains", today, on the other hand, it is up to us to *close* what little is left, *the very last reserves, closing it not for real friends, but for everyone and anyone connected with engineering and catering.*'
'You mean conservation.'
'Conservation is definitely an evil, as far as unspoiled Nature is concerned, just as a vaccine is in relation to real health. But we no longer have this choice. Now only conservation can save us from Die Norwegische Apparatlandschaft. I express myself in German.'

and later in the same article:

'But the people —?'

'... there's obviously no stopping them. But the day will come when they'll stop of their own accord. Today the Norwegian people weigh 220,000 tons. The nation's only common goal is to increase, to double, to multiply manyfold the number of flesh-bearers. The god of the age is called Multiplier. He is omnipotent and omnipresent. He guarantees that six times five are thirty, whether we are dealing with dirt or daffodils. Every single planned incubator is a temple raised in his honour. Terrace houses with terrace people, — blocks of flats with blocks of people — mass production of efficient nobodies. In the world of arithmetic no-one asks what the figure means.'

The article concludes as follows:

'I belong to a dying type of human being. That is why I say "Farewell, Norway". It's already in alien hands.'

Although, many might have a great deal of sympathy for Zapffe's delight in Nature, nature mysticism is alien to us as social scientists. As already mentioned, there are good grounds for assuming that economies, abandoned to the unhampered interplay of the forces involved, will generate pollution, and other forms of damage to the environment, to an excessive extent. As economists, however, it seems clearly wrong to believe that the answer to this mistaken organisation of our society is to lay the country wide open to the free play of natural forces.

An important contribution to the public debate about scarcity problems and pollution was the book *The limits to growth* written by a team of scientists (Meadows et al., 1972). In contrast to the smooth, neo-classical growth paths towards everlasting steady-states appearing in economic textbooks, this book gave a straightforward big-bang forecast. The options for the world economy were either to run out of natural resources or to be heavily polluted. Of course, a third alternative was specified in which all pollutants were recycled and where the utilisation of natural resources was stretched out in time to an extent not inherent in the present behaviour of the market economies. The conclusions and the reasoning — or the lack of reasoning — have been criticised by many. In his article 'World dynamics: measurement without data' W.D. Nordhaus pointed to the absence of empirical evidence for most of the conclusions drawn in the book.

Doomsday prophecies of this kind are nothing new. At the turn of this century prominent members of the Conservation Movement in the USA held the view that US might run out of natural resources such as coal and timber. They advocated public intervention in the market, first to prevent the exhaustion of natural resources, but later public intervention was given a wider justification. Henry George argued that public expenses should be financed by the scarcity rent occurring in the sectors where natural resources were input in the production processes.

From a professional point of view more important contributions came 100 years earlier, at the beginning of the nineteenth century. The two seminal contributors were Thomas Malthus and David Ricardo. A focal assumption of Malthus is that access to natural resources is given and limited. Natural resources are assumed to have a single quality. Up to the point at which the entire resources base has been exploited, the economy will grow, as though no limits existed to the supply of natural resources. After this point has been reached production will grow, but the growth rate will be reduced, and production will finally decline.

Ricardo assumed that access to natural resources was unlimited, but that the quality differed. He further assumed that the best resources qualitatively would be the first to be used, followed by the second best, and so on. This implies in the first place that society possesses knowledge that makes it possible to rank natural resources according to quality, and in the second place that society actually behaves in such a way that the best resources are used first. Given this, production is characterised by decreasing turns of scale, as indicated by Malthus. The difference between them is that decreasing returns will occur far more gradually in accordance with Ricardo's assumption than is the case with Malthus.

A work published in the postwar period, which is indispensable reading in this connection, is *Scarcity and Growth* (H.J. Barnett and C. Morse, 1963). Barnett and Morse set out to test Ricardo's hypotheses. Ricardo's hypotheses, as reproduced above, must be modified somewhat when the opportunity for technical progress exists. Barnett and Morse distinguish between two modified hypotheses. The most stringent hypothesis states that unit costs in extractive industries, such as

7

agriculture, forestry and mining, will increase over time, despite technical development. The weaker hypothesis states that unit costs in the extractive industries, compared with unit costs in other industries, will increase over time. Changes in unit costs are interpreted as an approximation to changes in the quality of natural resources. Barnett and Morse apply statistics from the USA covering the period 1870-1957. Both hypotheses were rejected, though with certain exceptions as far as forestry was concerned. Technical development has counteracted the trend towards decreasing returns of scale in extractive industries.

Although their approach and results can be questioned, it is fairly clear that the calculations of Barnett and Morse, as well as those of others, cast doubt on the validity of armchair speculations. This does not mean that we should close our eyes and leave economies to themselves. There is good reason to recall what one of the leading economists of this century, A.C. Pigou, declared in *The economics of welfare* (1920):

'But there is wide agreement that the State should protect the interests of the future in some degree against the effects of our irrational discounting and of our preference for ourselves over our descendants. The whole movement for 'conservation' in the United States is based on this conviction. It is the clear duty of Government, which is the trustee for unborn generations as well as for its present citizens, to watch over and, if need be, by legislative enactment to defend the exhaustible natural resources of the country from rash and reckless spoliation.'

However, the originator of today's countless studies on the exploitation of natural resources is not Pigou, but Harold Hotelling, who started his pioneering work *The economics of exhaustible resources* (1931) with the following words:

'Contemplation of the world's disappearing supplies of minerals, forests and other exhaustible assets has led to demands for regulation of their exploitation. The feeling that these products are now too cheap for the good of future generations, that they are being selfishly exploited at too rapid a rate, and that in consequence of their excessive cheapness they are being produced and consumed wastefully has given rise to the conservation movement. The method ordinarily proposed to stop the wholesale devastation of irreplaceable natural resources, or of natural resources replaceable only with difficulty and long delay, is to forbid production at certain times and in certain regions or to hamper production by insisting that obsolete and inefficient methods be continued. The prohibitions against oil and mineral developments and cutting timber on certain government lands have this justification, as have also closed seasons for fish and game and statutes forbidding certain highly efficient means of catching fish. Taxation would be a more economic method than publicly ordained inefficiency in the case of purely commercial activities such as mining and fishing for profit, if not also for sport fishing.'

Our work has been inspired by Hotelling's formulations of the problem in 1931. In common with Hotelling we are on the lookout for means that would prove more effective than direct control, in order to realise aims associated with the exploitation of natural resources. One reason why nearly 40 years were to elapse before economists once again showed interest in the problems posed by Hotelling might be the same as that which resulted in nearly 30 years passing before economists applied Frank P. Ramsey's pioneering work on optimal economic growth (1928), namely, that the economic development in the years that followed did little or nothing to stimulate thinking about long-term growth, optimal growth and growth limits. The 1930s were characterised by interrupted growth and preparations for war, events which increase perspective shortening. The future, on such occasions, counts for less than previously. The first postwar years, moreover, as far as economists were concerned, were marked by more short-term problems. One reason for this, of course, is that the Keynesian wave was almost nipped in the bud as a result of the outbreak of war in 1939. It was therefore reasonable to expect that to some extent postwar literature would bear the hallmark of unsolved and unpublished ideas in this field.

The interest in long-term resource problems which could be said almost to have exploded among economists around 1970 can be interpreted as a reaction to a long postwar period with hardly any break in growth. The breaks in growth which we experienced in the 1970s and 1980s will not detract from the interest in long-term resource problems to the same degree as in the 1930s since breaks in growth are to some extent tied to the rate of oil extraction and changes in this rate over time. Another explanation is that, thanks to a spread of knowledge on optimal control theories in the 1960s, economists now had a tool particularly well adapted to the analysis of long-term problems.

Pollution is not, of course, a new phenomenon. There are plenty of gloomy and harrowing descriptions of polluted immediate environments in older literature, such as Dickens' description of London in the nineteenth century. An early Norwegian contribution is provided by Ibsen's *An enemy of the people*, a drama that is spun around contemporary pollution problems. What is new in the present-day situation is the damage inflicted on the more distant natural environment, and one-time free goods such as water and air. Pollution problems now possess a potentially global character. It is pointed out that general changes in climate are possible. The earth's surface temperature may fall owing to the reflection of solar energy by pollution in the form of particles in the upper atmosphere, and because of increased reflection of energy from the surface of the sea caused by oil pollution. It is also possible that the earth's surface temperature might *rise* as a result of an increasing carbon dioxide content in the air; this might produce a greenhouse effect. The increasing amount of carbon dioxide is due to the increase in the use of fossil fuels. The actual thermal effect of this use of energy might also result in a rise in temperature. Air pollution in the lower atmosphere contributes to reducing the loss of heat from the earth. In Nordhaus' work (1977) the conclusion is reached, however, that remedying this would not cost a great deal in the form of a reallocation of resources.

It is also pointed out that the world's oceans could effectively be 'poisoned' by pollution so that, for example, the production of fish may be reduced, or there may even be a reduction of the ocean's transformation of carbon dioxide into oxygen in the carbon cycle.

External effects associated with pollution were first developed and elaborated by Pigou (*The economics of welfare*, 1920). By external effects we mean a situation in which factors are present in consumption and/or production that affect results over which the various units themselves have no direct control. The effects can only be changed by changes in the use of resources. Pigou's classical example was smoke from a

factory that dirtied washing hanging out to dry in the neighbourhood. Pigou revealed his interest in the empirical aspect by
referring to actual investigations. In 1918 an investigation was
undertaken by the Manchester Air Pollution Advisory Board on
laundry costs in Manchester compared with a clean town.
Weekly laundry costs of 100 comparable workers' families in
each town showed an extra expense of $7^1/2$ pence per household. The total extra cost per annum were estimated at
£290,000. Pigou pointed out that this might be a highly conservative estimate, as the expenses incurred by middle-class households had been equated with those of working-class households.

The person who definitely introduced pollution into the
analyses of economists, in such a way that pollution was
regarded as more than merely an exotic example of external
effects, was Allen V. Kneese. In 1969 together with R. Ayres he
wrote a pioneering article in this field.

An important contribution in the field of pollution is Rachel
Carson's *Silent Spring* (1962), in which she deals with the sort of
pollution caused by pesticides. These, however, are merely
examples of a large category of pollutions, namely, those associated with discharges in the past. This may involve substances
accumulating in the natural environment (pesticides and heavy
metals) or they may be of such a kind that the damage inflicted
on animals and humans is due to the fact that at some time in
the past they were exposed to a constant flow of pollution. Continuous noise pollution may result in chronic damage to hearing,
as well as stress. Air pollution may produce chronic bronchial
complaints. In our opinion these dynamic pollution problems
represent the most serious ones, both with regard to the nature
of the damage inflicted as well as to the scope. Furthermore,
this type of pollution involves far more complicated control
problems. An early Norwegian contribution in this field is
Trygve Haavelmo's article 'The pollution problem considered
from the socio-scientific point of view' (1971, in Norwegian).

Pollution caused by physical accumulation or load over a
long period are examples of irreversible encroachment on the
natural environment. Tolerance thresholds in the environment
and in human beings can be exceeded. Park-like tracts of old
forest may be cut down in order to accommodate a motorway;
the Arc de Triomphe may be flattened in order to allow for the
building of a car park, and so on. In order to decide whether
irreversible changes of this kind are to be undertaken, we

11

require a tool capable of tackling not only problems of irreversibility, but also uncertainty, since our information today about the wishes and opportunities of future generations is uncertain. These problems have been dealt with within the sphere of economic theory, *inter alia*, by Henry (1974), Arrow and Fisher (1974) and more recently in Fisher and Hanemann (1983).

REFERENCES AND FURTHER READING

Arrow, K.J. and A.C. Fisher (1974): 'Environmental preservation, uncertainty and irreversibility', *Quarterly Journal of Economics*, LXXXV III, 88, 312-19

Ayres, R.U. and A.V. Kneese (1969): 'Production consumption and externalities', *American Economic Review*, 59 (7), 282-97

Barnett, H. and C. Morse (1963): *Scarcity and growth, the economics of natural resources availability*, Johns Hopkins Press, Baltimore

Beckerman, W. (1974): *In Defense of Economic Growth*, Jonathan Cape, London

Carson, R. (1962): *Silent Spring*, Fawcett Publications, Inc., Greenwich, Connecticut

Fisher, A.C. and W.M. Hanemann (1983): 'Endangered species: the economics of irreversible damage', University of California, Department of Agricultural and Resource Economics, Working Paper no. 329, Berkeley

Fisher, A.C. (1981): *Resource and environmental economics*, Cambridge Surveys of Economic Literature, Cambridge University Press

Henry, C. (1974): 'Option values in the economics of irreplaceable assets', *Review of Economic Studies*, 89-104

Hotelling, H. (1931): 'The economics of exhaustible resources', *Journal of Political Economy*, 39, 137-75

Kneese, A.V., Ayres, R.U. and d'Arge, R.C. (1970): *Economics and the environment: a materials balance approach'*, Johns Hopkins Press, Baltimore

Kneese, A.V. and J.L. Sweeny (1985): *Handbook of natural resource and energy economics*, vol. I, North-Holland, Amsterdam

Kneese, A.V. (1977): *Economics and the environment*, Penguin Books, Harmondsworth

Malthus, T. (1798): *An essay on population*, London

Meadows, D. *et al.* (1972): *The Limits to Growth*, Universe Books, New York

Mills, E.S. (1986): *The economics of environmental quality*, 2nd edn, W.W. Norton & Co., New York

Nordhaus, W.D. (1973): 'World dynamics: measurement without data', *Economic Journal*, 83, 1156-83

Nordhaus, W.D. (1977): 'Economic growth and climate: the carbon dioxide problem', *American Economic Review*, 67, 341-6

Næss, A. (1975): '*Økosofi T*' in *Økologi og økofilosofi*, edited by P. Hofseth and A. Vinje, 150-63, Gyldendal Norsk Forlag, Oslo

Pigou, A.C. (1920): *The economics of welfare*, Macmillan, London

Ricardo, D. (1817): *Principles of political economy and taxation*, J.M. Dent, London, 1965

Smith, V.K. (1979): 'Natural resource scarcity: a statistical analysis', *Review of Economics and Statistics*, 61, 423-7

Zapffe, P.W. (1969): *Barske Glæder*, Gyldendal Norsk Forlag, Oslo

Part II

Pollution

3

Concepts

3.1 INTRODUCTION

The purpose of Part II is to consider a limited number of pollution problems in an economic context. In order to obtain an overall view of fundamental problems, it has been considered useful to introduce simplifications of the relationships in the natural environment and of relationships between the natural environment and economic activities. The price for a necessary pedagogical simplification of this kind is that a number of concrete phenomena cannot be recognised in the aggregate concepts used.

Economic activity, production and consumption, influences the natural environment in three fundamental ways:

(1) by the occupation of space;
(2) by the extraction of raw materials, such as minerals, oil, fish, etc.;
(3) by the discharge of residuals that find no further use in economic activities.

Problems relating to the last mentioned point will be discussed in Part II. Discharge of residuals from production and consumption is the cause of the rapidly growing pollution of water, air and earth in countries with a large consumption of raw materials and energy.

In order to analyse the relationship between economic activities and pollution, it will be convenient to distinguish between the following three steps:

(1) the relationship between economic activities and the discharge of residuals;
(2) changes in Nature due to these discharges;
(3) social costs related to these changes in the natural environment.

3.2 DISCHARGE OF RESIDUALS

Discharge of residuals is a fundamental feature of economic activities. The throughput of materials is illustrated in Figure 3.1. Materials of substances used in economic activities are incapable of disappearing in a physical sense. We can draw up a material balance for the economy. The amounts of materials or substances extracted from Nature must either remain in the economic cycle, or be discharged and return once again to the natural environment.

From now on we shall use the term *residuals* to indicate materials and products returned to Nature after having been utilised in the economic cycle. Measured in terms of weight, the amount of residuals from an economic activity is equal to the amount of materials input (including oxygen, water, etc.) minus the amount of end-products of that activity and minus the net change in stock and amount of real capital. Rough calculations for the US economy arrive at an input of about 12 tons of goods per person in 1965, including net imports. It is assumed that approximately 10 to 15 per cent of this is tied up in the form of real capital, mainly buildings.

Figure 3.1: The throughput of materials in the economy

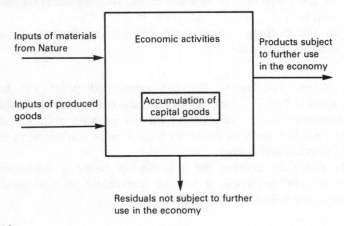

18

In the case, for example, of the consumption of food, clothes, dwellings, cars, etc. the physical objects may be regarded as transmitters of services such as comfort, heat, transport, etc. Consumer goods are either transformed directly into residuals or stored in the form of durable consumer goods. In principle, changes in the weights of consumers themselves should also be taken into consideration.

Physical wear and tear on various forms of real capital also generate residuals, both by the actual wear and tear itself and in the form of scrap when replacing capital goods, for example, car wrecks.

Residuals can be classified according to various criteria; for example, according to physical conditions and chemical composition or according to the effects of the discharges. The main types of residual are:

material residuals:
   solid
   fluid
   gaseous
energy residuals:
   heat
   noise (transformed to heat)
   radiation

3.3 THE EFFECT OF RESIDUALS ON THE NATURAL ENVIRONMENT

So far Nature or the natural environment has been used as a collective term. For the sake of convenience the natural environment can be divided into recipients. The main types of recipient are air, land and water. Recipients may be more or less naturally delineated, for example, a lake, a river, a particular area of land, the air space above a city and so forth. A recipient has a certain normal condition, characterised, *inter alia*, by certain forms of plant and animal life included in an ecological system. We do not intend here to give a more detailed natural scientific description of recipients, but will concentrate instead on the minimum of insight necessary to understand the relationship between economic activity and changes in Nature.

It is presumed that the state of recipients can be described in terms of certain measurable conditions, which we may call environmental indicators. Examples of such indicators are:

the oxygen content per volume-unit of water;
the quantity of fish in a lake;
the amount of bacteria per volume-unit of water;
the acidity of water measured in pH values;
visibility depths of water;
the quantity of algae in water;
concentrations of sulphur compounds, dust, carbon monoxide, mercury, lead and nitrogen compounds in the air;
the number of birds within a certain area;
the probability of contracting ailments such as bronchitis, asthma, lung cancer and thrombosis.

Environmental indicators of this kind are often correlated, partly because the discharge of residuals may influence several of the indicators used to describe the state of the recipient, partly because the environmental indicators are interrelated through the internal structure of the ecosystem and finally because different pollutants are contained in the same discharge.

Under constant external conditions balance will occur in the recipient between the various environmental indicators. Any alterations in the factors involved in this balance will displace the equilibrium. The magnitude of this displacement will depend on the influence exercised by the factor in question upon the equilibrium, and will often be difficult to forecast, even if the direction of the displacement is known.

A number of factors included in the system take the form of a more or less continuous supply of materials, for instance oxygen, water and the dissolved nutrients. If the continuous supply of important materials is increased or reduced, some time will elapse before equilibrium is re-established. The observed environmental indicators can often show a time variation, as illustrated in Figure 3.2.

When the supply of residuals suddenly increases, some organisms may after a while be reduced in number, or possibly die out, while others will increase in number. If the external change is not excessive, a new equilibrium will be established. However, it is also possible to imagine that if the sudden increase is sufficiently great, then the decrease in the level of the

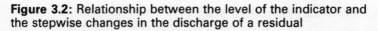

Figure 3.2: Relationship between the level of the indicator and the stepwise changes in the discharge of a residual

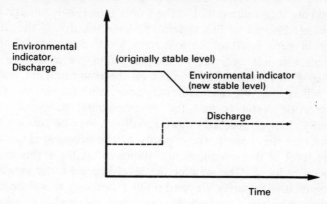

indicator may be so considerable as to be reduced to almost zero. This illustrates the thresholds for discharge of residuals. If such thresholds are exceeded, the environmental indicators (for example, the oxygen content in a lake) may after a while be reduced to almost zero. The transitional period itself, too, may also be influenced by the magnitude of the load. The greater the load, the faster may be the transition from the original positive level to zero.

If the external load increases steadily but slowly, in such a way that the system remains approximately in equilibrium the whole time, the system might develop as shown in Figure 3.3. Initially, small amounts of residual are discharged: the recipient is capable

Figure 3.3: Possible relationship between environmental indicator and discharge of residuals

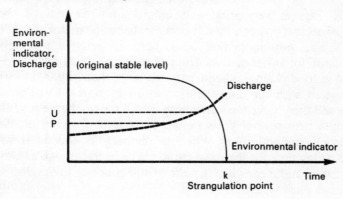

21

of assimilating them. Most of the environmental indicators remain unchanged, and normal conditions prevail. The amounts discharged are then increased, but the whole time slowly enough to ensure equilibrium in the system. This means that if the discharge is kept constant at a certain level, the state of the recipient will not noticeably change. When the discharge is equal to P in Figure 3.3, the effect on the environmental indicator will appear. When the discharge continues to increase, the characteristic result is that the environmental indicator will gradually change more and more rapidly. It may be realistic to assume that the assimilative capacity of the recipient depends on the level of the environmental indicator. After P this level has been reduced. The smaller the level becomes, the smaller the assimilative capacity. Beyond point P one may therefore say that the assimilative capacity is more and more reduced, while at the same time the discharge increases over time. For this reason the environmental indicator will plunge more and more steeply towards strangulation point.

What might happen if the discharge stops completely, involving, for instance, a reduction from level U to zero? If the discharge stops, and provided the balance in the system is not destroyed, the recipient may after a time revert to its normal state.

The time that ensues before the system returns to its normal condition depends, however, on the speed of circulation through the system and its character. If the speed of circulation is determined by the ventilation in the system in the form of wind or water circulation, a relatively brief period of time may elapse before the recipient returns to its original state (with the exception of such irreversible changes that might have taken place, say, in vegetation and animal life). Several of the pollutants that do not normally occur in the recipient, for example DDT, and substances that occur naturally only in very tiny amounts, for instance mercury and cadmium, may be assimilated in food chains and concentrated more than 100,000 times compared with the concentration in air or water. Circulation time will then be very considerable, and if the discharge is small it could take several years before the effects show up on the environmental indicators. When the discharge is stopped, it will take a long time for the recipient to purify itself through natural processes, and meanwhile the effects will continue to develop in the ecological system.

A few examples will suffice to illustrate this. If, for example, a certain amount of sulphur dioxide and dust is discharged into the air over a city, the air will be cleansed by wind and rain. A dynamic equilibrium will then be established between input and output of residuals, changing with the weather conditions.

The convenient time-unit to work with depends on how the actual damage occurs. Acute health effects have to be studied within, for example, a day, while long-term corrosion effects can be observed on a yearly basis. Decomposition of organic waste in water will first produce a reduced oxygen content in the water, and then a higher oxygen content again, once decomposition has been completed (see Figure 3.4). If an oxygen content that is lower than normal does not temporarily produce any deleterious effects, it may be natural to operate with a time-unit for the registration of discharge long enough for the fluctuation in the oxygen content not to register.

In this case, however, the amount of sunlight will also influence the photosynthesis of algae etc. living in the water and producing oxygen.

This process produces oxygen during the day, but the decay of dead plants consumes oxygen both night and day. Run-offs from silos into rivers promote a rapid growth of algae, with a high production of oxygen during the day. This may lure salmon to enter the rivers, but during the night the oxygen content sinks to the point where salmon are suffocated. Several instances of ill effects of this nature have been observed in Norway as well as in other countries, and variations within the 24-hour span are here obviously of importance.

DDT has a decomposition time of about 25 years, and the accumulation in the biological food chains takes place much more rapidly than decomposition. For this reason, in a polluted area the ill effects will be felt a long time after the discharge of pollutants has ceased.

From these examples it will appear that in several cases the level of the environmental indicators does not depend solely on discharge at the same time, but also on discharge in earlier periods. The amounts of pollutants can accumulate. Examples of such residuals include heavy metals, chemicals used in agriculture and forestry, and plastic substances. These forms of pollutant are of special interest in the analysis of economic growth, and some of the problems will be discussed in greater detail in Chapter 8.

23

An irreversible process can, for example, be illustrated in Figure 3.3 by assuming that, once the point of strangulation has been reached, the environmental indicator cannot revert to normal values with a cessation of discharge. It is then assumed that the system has been destroyed. Even if the discharge of pollutants is reduced or completely stopped, the recipient cannot through natural processes return to its normal state in a case of this kind. Irreversible developments will be discussed in greater detail in Chapter 9.

Ideally, it might have been supposed that economists could obtain suitable relationships from the natural sciences in question, subsequently employing them in economic analysis. In practice, this is not quite so simple: one reason could be the one indicated by Kenneth Boulding, namely, that the natural scientists are highly expert in dealing with the very detailed and the global, but strangely ignorant about systems in between. The simplifications of economists may appear strange and downright unsuitable to natural scientists.

With regard to 'ecological' models utilised in economic analysis, the best established relationship is the one between continuous discharge of decomposable organic material into water and the oxygen content of water. For example, in a river the level of oxygen will follow a typical pattern of decreasing level followed by a rise to the original level downstream, as illustrated

Figure 3.4: The effects of discharge of organic waste on oxygen concentrations downstream

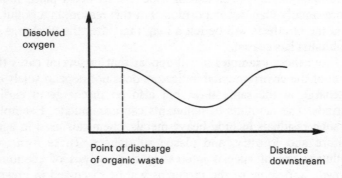

Figure 3.5: Discharge of particles to air and concentrations downwind

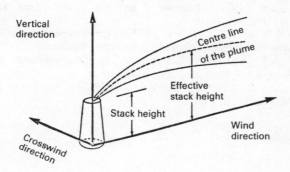

in Figure 3.4. The process can be simulated by a system of differential equations, the so-called Streeter-Phelps equations, introduced as far back as the 1920s. The solution of a system of this kind makes it possible to estimate a set of coefficients for an equilibrium that gives the relationship between *repeated* discharges and the level of oxygen at different points down river.

With regard to the diffusion of particles in the air, comprehensive and complicated models exist which are operational. These are based on the simplification illustrated in Figure 3.5. The plume from the point-source, calculated in terms of the effective stack height, is presumed to exhibit a Gaussian distribution of concentrations in the cross-plume and vertical dimensions. When the standard deviations in these normal distributions are known, together with the average wind speed and direction, the concentration of particles can be calculated for selected points around the point-source.

Ecological models now exist for aquatic phenomena, such as the growth of algae, as well as models for the formation of photochemical smog in the air, which can be used together with economic models (see Russell (1975)).

REFERENCES AND FURTHER READING

Ayres, R.V. and A.V. Kneese (1969): 'Production, consumption and externalities', *American Economic Review*, 59, 282-97

Kneese, A.V. (1977): *Economics and the Environment*, Penguin Books, Harmondsworth

Kneese, A.V. and B.T. Bower (1979): *Environmental quality and residuals managment*, Johns Hopkins Press, Baltimore

Kneese, A.V. and B.T. Bower (1972): *Environmental Quality Analyses*, Johns Hopkins Press, Baltimore

Russell, C.S. (ed.) (1975): *Ecological Modeling*, Resources for the Future, Washington, DC

4

The Main Problems

4.1 SERVICES OF THE NATURAL ENVIRONMENT

One approach to an economic analysis might be to consider the natural environment as a form of production capital: Nature 'produces' certain goods and services. In order to arrive at the principal problems involved, the services provided by Nature can be divided into three main categories:

(1) waste disposal services
(2) extraction services
(3) amenity services

Discharge of residuals from economic activities is countered by a waste disposal service provided by Nature. The exploitation of substances existing in the natural environment, such as minerals, oil, forest, fish, water, oxygen, etc. is countered by extraction services. Amenity services, here intended in a very wide and comprehensive sense, may include everything from open-air activities, angling, bathing, etc. to aesthetic experiences, and the view that Nature possesses an innate value in itself.

A distinction can often be made between waste disposal services on the one hand and extraction and amenity services on the other. The core of pollution problems is that a conflict exists between these two main types of services: the use of the waste disposal services may have a negative effect on the scope or quality of extraction and amenity services. Examples of types of negative effects could include the following:

the quality of the open-air services declines;

the incidence of bronchial ailments, various forms of cancer, etc. increases;

the production of a given set of goods and services will require more resources;

the maintenance of a given level of consumption will require more resources of the consumer (for example, more detergents in order to keep clothes clean)

A possible connection between types of services is shown in Figure 4.1. Extracting activities may also have negative effects on amenity services, e.g. forestry, mining, hydropower development.

In Figure 4.2 a possible connection between the discharge of residuals and negative effects is illustrated as a reduction in extraction and amenity services. It is important to note that while environmental indicators (Section 3.3) can be measured objectively, the negative effect will often be merely subjectively measurable. This is especially clear in the case of amenity services, as these are seldom subject to evaluation, as is the case with goods offered for sale in a market. The quantification of negative effects may vary from individual to individual; some people, too, may in principle reject the idea of a quantification, a view that will be commented on in Chapter 5.

The conclusion to the problem of quantification is that it is not always possible to arrive at a definte numerical evaluation of negative effects. It may therefore prove necessary to base any concrete measures aimed at influencing discharge on an assess-

Figure 4.1: Possible connection between extraction and amenity services and waste disposal services

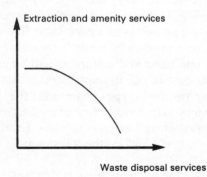

Extraction and amenity services

Waste disposal services

Figure 4.2: Possible connection between discharge of residuals and the reduction in extraction and amenity services

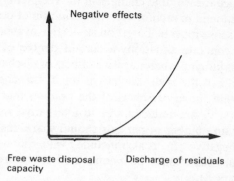

ment of the evaluation of various groups. Conflicting interests of this nature will be discussed in section 4.4.

If a recipient possesses a self-cleaning ability for a certain type of pollutant, a certain amount of residuals can be discharged before extraction and amenity services are negatively affected. In addition to natural self-purification the situation may also exist in which a negative reaction is not registered, even though the environmental indicators are to some extent affected. In cases where a recipient is capable of absorbing residuals without negative effects occurring we can speak of the recipient possessing a free waste disposal capacity.

There is no conflict between waste disposal services on the one hand and extraction and amenity services on the other for the discharge of residuals within this limit. This presupposes a complete overall view of negative effects, of a long-term nature as well. Certain cases may also be conceived in which a certain amount of residual is evaluated as a positive improvement of the recipient services, for example, certain fertilising effects.

4.2 POLLUTION IN THE ECONOMIC SENSE

Pollution in the economic sense can be defined with the aid of the service concepts introduced in Section 4.1. That pollution exists means that discharges of residuals (that is, the use of waste disposal services) reduce the quality and/or scope of extraction and/or amenity services. When discharges do not produce negative effects, that is, when they are within the free

29

waste disposal capacity illustrated in Figure 4.2, residuals will not be designated pollutants in the economic sense, even though they produce measurable changes in the recipient.

As mentioned in section 4.1, the definition of pollution in the economic sense must be based on subjective evaluation. For this reason in concrete situations a certain degree of arbitrariness will exist in judging when negative effects of discharge occur. At the one end of the scale discharges might be called pollutants, provided one single person is of the opinion that the environmental services are reduced. As an alternative we might insist that a certain number of persons should share this view. Evaluations of negative effects also include evaluating which environmental qualities should be bequeathed to future generations (see Chapter 8).

The use of Nature's waste disposal services involves a certain degree of expense in the socio-economic sense once the free capacity for waste disposal is exceeded. 'Natural capital' has an alternative use in the 'production' of extraction and amenity services. The natural environment can be conceived of as an economic resource on a par with what we usually describe as economic resources. The use of scarce natural resources should then be subject to the same demands of efficiency that apply to the exploitation of other resources in supply.

The fact that pollution is regarded as a problem must in the first place mean that the free waste disposal capacity has been exceeded, and in the second place that the relationship between the waste disposal services and the extraction and amenity services is not satisfactory.

Within an economic context a solution to pollution problems involves realising the 'correct' relationship between the waste disposal services and extraction and amenity services, that is, realising a particular point on the curve in Figure 4.1. By 'correct' relation is meant that by small changes, from a particular point on the curve in Figure 4.1, the negative effects of a reduction in the extraction and amenity services are to be precisely balanced by the positive effects of increased waste disposal services. The utility of waste disposal services is expressed by the result of the economic activities that utilise these services.

4.3 POLLUTION AND INDIRECT EFFECTS IN THE ECONOMY

If scarce resources are to be utilised effectively, due regard must be taken of all alternative uses. The alternative to increasing the use of Nature's waste disposal services is to increase the use of the amenity and extraction services. It is not to be expected that the waste disposal capacity will be correctly utilised, if the decision-making bodies fail to take alternative use into consideration.

Providing mechanisms that ensure effective utilisation of resources is one of the main tasks of economic theory. In a free market the price mechanism plays a dominant role in the allocation of resources for various purposes. The social costs of utilising a resource, however, must be reflected in the price of that resource, if the price system is to function effectively.

The question that may arise in this context is whether the negative effects of pollution are reflected to a sufficient degree in the production costs of the goods and services responsible for this pollution. Statements to the effect that pollution exists on an excessive scale may be interpreted to mean that the negative effects have not to a sufficient extent been absorbed in the market system.

The fact that economic activity creates negative effects that are not reflected in market prices is not a new problem in economic theory. The concept 'external effects' was introduced in the previous century by Alfred Marshall to describe both positive and negative indirect effects. This term was used in economic analysis to discover which changes would have to be made in a system that falls short with regard to effective exploitation of its resources.

The principal economic solution to problems involving indirect effects as applied to problems of pollution may briefly be said to involve the following:

— finding the prices for waste disposal services corresponding to the social marginal cost, measured in terms of alternative use of the natural environment to supply us with extraction and amenity services;
— establishing systems whereby the decision-making bodies take these prices into account.

In Chapter 5 we shall consider the sort of systems that might be established.

If this is carried out we may say that the negative effects of pollution are introduced into the economic system in line with other magnitudes which are usually described as economic.

4.4 CONFLICTING INTERESTS

There are two main types of conflicting interests: one of these involves the environmental qualities to be selected. As pointed out in section 4.1, there is unlikely to be general agreement on, for example, to what extent a particular watercourse should be loaded with pollutants, etc. This conflict also involves evaluations over a period of time. It is also important to decide what kind of waste disposal capacity the next generation is to inherit. This point will be discussed in Chapter 8. Low-income categories may be less concerned in sacrificing some traditional form of consumption now, with a view to improving the environment in the future, than would be the case with higher-income categories.

The following is an example of a conflict situation of this kind. Let us assume that we have two categories of individuals in the community; one may have a far lower consumption level *per capita* than the other category, while at the same time the latter consists of far fewer individuals. The consumption of the great majority of the individuals may be assumed to involve mass production, industrialised products, etc. The consumption of the higher-income group has a much higher proportion of services. Let us also assume that mass production is responsible for future pollution. One fact is quite clear: if pollution is to be reduced, as desired most strongly by the higher-income group, little can be done by reducing *their* consumption. In the first place direct reduction in pollution would be slight, if the consumption of the higher-income group were reduced. In the second place the resources made available by a reduction of the consumption of this group would possibly only produce small effects if these resources were utilised for some kind of purification activity.

The other kind of conflicting interest involves the distribution of environmental services and the goods and services produced. This is not a new conflict peculiar to environmental protection measures; but it is obvious that measures of this kind may have distributional effects that will produce greater or lesser differences in the standard of living. For example, certain recreation services can only be enjoyed by consumers with the input of produced goods and services, such as transport to recreation areas, a boat, fishing rod, country cottage, and the like. The ability to make an input of this kind will depend on income. For this reason a contribution to preservation of the environment, as far as recreational services of this kind are concerned, would be of relatively greatest use to the higher-income groups.

Differences in income and consumption levels may therefore comprise an important factor to be taken into account in an analysis of pollution problems and in the implementing of measures.

4.5 INTERNATIONAL ASPECTS

From the point of view of the geographical location of the recipients and sources of discharge, international pollution problems derive from two circumstances:

(1) foreign sources pollute national recipients via natural transport by air and water;
(2) discharges from sources in one or several countries pollute international recipients that provide services to one or more countries.

An example of the first type is the transport of acid components from the Continent and the British Isles to Scandinavia. These fall out with precipitation and may reduce stocks of trout in lakes and rivers in south Norway. An example of the other kind is the pollution of international seaways such as the Baltic and the Mediterranean, the North Sea, etc.

In our part of the world complaints are raised about the increasing pollution of stretches of sea such as the North Sea and the Atlantic, and acid rain over Scandinavia. If these pollution phenomena are to be tackled, something must be done at source. Collective decisions must be taken by groups which, on historical grounds, are not used to making decisions of this nature.

Up till now global pollution phenomena have been caused by

the mass production of the industrialised countries and consequently their increasing material standard of living. The consumption level *per capita* in the developing countries is far lower than in the industrial countries. Let us suppose that the developing countries achieved a consumption level per capita approximating the *per capita* consumption level in the industrial countries. Owing to the large number of individuals in the developing countries this might entail pollution problems of far greater dimensions than those that have been the subject of general debate. This raises several problems:

(a) Populations of the developing countries would hardly be willing to renounce increased consumption, if this were possible. In their case protection of the environment, to the extent that this would reduce their material conditions, would prove a luxury.

(b) It would probably be of little help if the industrial countries sacrificed some of their per capita consumption level with a view to containing their pollution problems within the limits reached so far, and at the same time gave the developing countries an opportunity of achieving a higher per capita consumption.

(c) Even supposing action of this kind were to have some effect, it would probably entail such considerable sacrifice that it might be politically difficult to carry out, almost irrespective of what political views one might hold.

(d) As a consequence of the above, if the global pollution problem is to maintain its present dimensions, and on the assumption that no drastic steps can be taken with a view to reducing the population growth in the developing countries, the *per capita* consumption in the developing countries must not be allowed to exceed its present level.

Or, to quote Professor Trygve Haavelmo ('The pollution problem considered from the socio-scientific point of view', 1971):

If we fail to achieve a drastic reduction in the population pressure, ideas of any policy of environment protection on a global scale are either mere talk, or are implicitly based on the assumptions of a world in which there are still a few *haves* and a great many *have nots.*

For this reason population growth is highly relevant in any long-term discussion of pollution. Official control of population growth should therefore have been one of the subjects discussed in this book; although this is a tremendously important problem, space does not permit a discussion of it, for which reason we shall merely stress its importance.

Problems of the pollution of international recipients may be more easily solved by international co-operation than foreign pollution of national recipients, in circumstances where the countries affected all derive some advantage from the contributions of the international recipient. The Oslo Convention on the prevention of ocean pollution by the dumping of waste oil from ships and aircraft is an example of this kind. Foreign pollution of national recipients has a bi- or multi-lateral character, and solutions to this depend on established rules aimed at settling disputes between the countries involved. A special problem arises in the case of imported goods which, when used, produce a greater degree of pollution in the form of DDT, PCD, etc. than corresponding Norwegian-produced goods.

We do not possess sufficient figures to establish the amount of residuals in our national recipients that originate from foreign sources. We have a sufficient basis, however, for arguing generally that the lion's share of residuals, except for SO_x, originates from national sources. For this reason there is no cause to desist from national measures aimed at preventing pollution due to production and consumption in Norway, while awaiting international agreement on international pollution.

In the case of a country with a substantial foreign trade, environmental protection measures may also have some bearing on the composition and scope of the exchange of goods with foreign countries. This is, of course, nothing negative, but something we desire. If our export industry is not to be burdened with the social costs involved in the discharge of residuals, this means that we are subsidising the foreign consumption of these goods by imposing an excessive burden on our natural environment. It has been maintained in certain quarters that countries trading with one another should adopt equally strict measures with regard to mutual competition. The following arguments against this can be mentioned. First, the natural environments of various countries may differ in their ability to receive residuals. Some countries may have rivers and stretches of coastline possessing a high degree of self-purification, other countries

35

may, for example, place greater reliance on lakes and rivers with a lesser degree of self-purification. Secondly, the inhabitants of various countries may have different preferences, *inter alia*, with regard to the exploitation of the natural environment, for example, for recreational purposes. Even though different countries might institute a policy of charges, so that indirect effects were incorporated in the market system and national evaluation could provide the basis for the magnitude of such charges, there are no grounds in the short run to assume that these charges would be equally great. In the long run, however, in the free, international exchange of goods there may be a tendency for environmental capital to be given the same 'reward' in the various countries, thereby involving equally large charges from one country to another.

One important point is that failure to do anything about factories and the like that cause pollution means continuing to subsidise the establishments concerned, for the simple reason that they are not induced to take into consideration the social costs involved in exploiting the natural environment. These costs *per se* are just as real as, for example, labour costs. The difference is that, unlike labour costs, none of these resources are essentially privately owned. However, the authorities, ideally speaking, should charge these costs to the market participants. Instead of saying that such charges cannot be introduced, one might say that we cannot afford this form of subsidy. The fact that some countries continue subsidising is no reason why one's own country should be compelled to do the same.

But in a situation in which a number of countries plan to enforce environmental protection measures, it is naturally an advantage to co-ordinate such measures in time. Measures may, for instance, involve costly adjustments in the export firms involved. Co-ordinating measures would prevent unnecessary adjustments. Reduced employment or the closing down of factories owing to environmental protection measures does not necessarily warrant special treatment on a permanent basis for export industry. Modification or adjustment of environmental protection measures during a transitional period may, of course, prove necessary. We shall deal with this aspect in Chapter 5.

4.6 THE POSSIBILITY OF REDUCING (CONTROLLING) POLLUTION

In working out meaningful measures information is necessary on a number of concrete facts and circumstances. Taking as our basis section 3.1, we may distinguish between the following four stages:

(1) discharge of residuals from production and consumption;
(2) natural reactions in recipients to the discharge of residuals, for example, dissipation, decomposition, transportation of residuals between recipients;
(3) effects on extraction and amenity services;
(4) society's evaluation of changes in these services.

In order to be able to draft concrete measures we must also be familiar with the possibilities that are available for influencing:

(1) discharge of polluting residuals from the various sources;
(2) what the recipient receives in the way of residuals;
(3) the natural processes taking place in recipients.

At every stage the possibility exists of organising things differently from what is the case at present. A highly important point as to environmental protection measures is to make use of the alternatives available in every single case.

The following possible choices exist in principle for affecting residuals from, for example, the manufacture of a good:

(a) reduced production, that is less use of materials and energy;
(b) a change in the composition of the input factors: for example, a transition from heavy to light fuel oils with a lesser sulphur content;
(c) a change in technology: for example, more complete combustion in engines, closed electrolysis furnaces, and the like;
(d) a change in products: for example, by cutting out phosphates in detergents;
(e) recycling or recirculation of residuals so that they are utilised once again in economic activities;
(f) treatment or purification of residuals before they are discharged to the recipients;

37

(g) relocation of activities to recipients with greater capacity for waste disposal;

(a) to (d) above involve internal organisation of an activity.

It is generally maintained that technical changes constitute the most important factor as far as reducing pollution on a long-term basis is concerned. Technical changes may effect a saving in raw materials, as is the case with recirculation and recycling. However, this is not always so. The addition of calcium to sulphuric gases in order to neutralise them before discharge might, for example, increase the consumption of raw materials. Our present type of economy is described as a throughput economy. Increased recirculation and recycling tend in the direction of a type of economy comparable to a closed system in a spaceship.

In addition to what can be done with the actual production processes, residuals can be dealt with before they are emitted to the recipients. The aim of purification is to give residuals the sort of treatment that will ensure a different physical or chemical form and/or disperse them to other recipients than those otherwise involved in discharge from the activity concerned.

Purification simply means that waste disposal services producing pollution are exchanged for other waste disposal services producing less harmful pollution. The varying waste disposal capacity of different recipients also offers an opportunity for reduced pollution by relocating activities. Whereas the advantages of external purification enjoyed by large plants involve concentration, better utilisation of waste disposal capacity by relocation may warrant spreading the sources. This, however, depends very largely on local conditions.

The potential for changes is considered from the point of view of a single activity. Changes on that sort of level, however, may have further effects owing to the mutual interdependence existing in an economy. Manufacturing units supply goods to other manufacturers and to consumers. The result of a successful ecological policy must be to change the composition of goods and services, including Nature's services.

In order to exercise economic control on the effect of any measure, it is necessary to have a frame of analysis embracing the mutual interdependence of the economy, so that stabilisation effects, the effects on distribution, changes in the relationship between consumption and investments, etc. can be studied. Mutual economic dependence between production units can be studied in input-output analysis. Extensions of economic models based on this sort of framework are discussed in Førsund (1985), Førsund and Strøm (1974), Kneese *et al.* (1970), and Leontief (1970).

The natural self-purification in the recipients can be influenced. The rate of decomposition may depend on the load and degree of concentration of residuals with regard to both volume and time. One example of assisting Nature's capacity for self-purification is the artificial infusion of oxygen in water by blowing in air, in order to hasten the breaking down of organic substances. The harmful effects of certain residuals can be neutralised by the addition of other substances, for example, lime in acid soil or acid water.

The link between the waste disposal services and the extraction and amenity services can also be changed by protecting anyone or anything exposed to their negative effects (human beings, materials, animals and plants) by, for example, additives in drinking water, air filters, double glazing against noise, etc. The natural services can also in certain cases be replaced by manufactured services. A swimming pool with purified water, for example, can provide for some people a very efficient substitute for a polluted lake.

Man's preferences with regard to environmental qualities are not dictated by Nature. We can be influenced to appreciate certain effects of residual discharge or technical encroachments on or violations of the natural environment. For instance, we might assume that certain polluted states of the natural environment, if they last long enough, are accepted as normal.

4.7 ACID RAIN AND FRESHWATER FISH

The detrimental effects of acid rain from Great Britain and the Continent on populations of trout in Norwegian lakes and rivers serve as a background for presenting an extremely stylised model making use of many of the main concepts introduced so far in Chapters 3 and 4. The model encompasses the following variables:

SO_x = sulphur oxides
X = amount of consumer goods ('stuff')
pH = indicator for acidity in water

$$F \quad = \text{ stock of fish}$$
$$W \quad = \text{ a measure of welfare}$$

Among these variables we have the following relations:

$$SO_X = a(X), \, a' > 0 \tag{4.1}$$
$$pH = b(SO_X), \, b' < 0 \tag{4.2}$$
$$F = c(pH), \, c' > 0 \tag{4.3}$$
$$W = d(F, X), \, d'_F > 0, \, d'_x > 0 \tag{4.4}$$

Relation (4.1) shows the generation of residuals, sulphur oxides, SO_X, by economic activity, consumption of stuff, X. Stuff is freely available in the economy. Increased consumption generates increased discharges of residuals ($a' > 0$). Relation (4.1) represents the first stage of the four stages in an economic analysis of pollution identified in section 4.6.

Relation (4.2) describes the natural reactions in water recipients when sulphur oxides are discharged. Time lags and transportations out of our model economy are disregarded. The acidity of the lakes increases (lower pH value, $b' < 0$) when discharges of sulphur oxides increase.

Relation (4.3) portrays the ecological consequences of increased acidity. The stock of fish is negatively affected ($c' > 0$).

The last stage in our classification of environmental economic analysis is the evaluation of the two goods in our eco-

Figure 4.3: Generation of residuals by economic activity

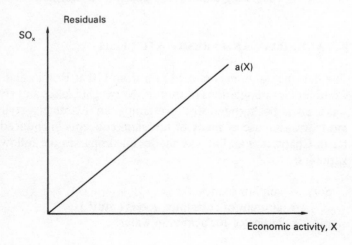

Figure 4.4: Discharge of residuals and environmental indicator

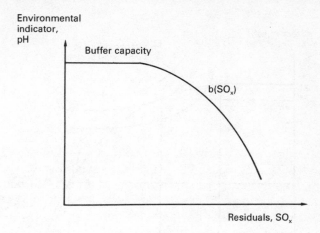

Figure 4.5: Stock of fish and environmental indicator

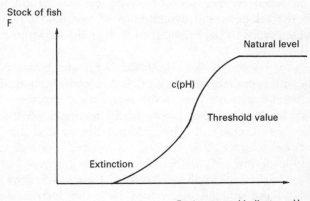

nomy: fish and stuff. The social preferences are expressed by the relation (4.4). Preference for fish can be interpreted both as preference for fish consumed and for fishing as a recreational activity. We are happier the more we get of both goods (d'_F, d'_x > 0).

The fundamental problem in our economy is that the more freely available stuff is eaten the less fish is available. A balance must be found between stuff and fish.

Figure 4.6: Social optimum of fish and stuff

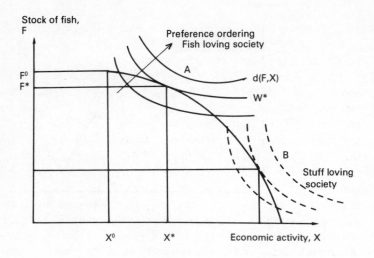

The social balance may be found by maximising welfare given the relations and assumptions of our model (4.1–4.4). Possible graphs of the relations (4.1–4.3) are shown in Figures 4.3–4.5.

Figure 4.3 shows that residuals, SO_X, are generated *pari passu* with economic activity modelled as consumption of stuff. The water recipients have some assimilative capacity, due to natural levels of calcium, before acidity increases when sulphur oxides are discharged. The fish population also has some capacity to tolerate acidity before it is reduced. For high enough acidity level (low pH) the fish die out as has happened in thousands of lakes in southern Norway. Figure 4.5 shows both threshold and strangulation values. Note that increased acidity is measured by a decrease in pH value, so Figure 4.5 is the mirror image of Figure 4.4.

At zero economic activity the fish population attains its natural level. As economic activity is gradually increased generation of residuals increase. But Nature has assimilative capacities within certain ranges both as regards increase in acidity of water recipients and as regard the tolerance of fish to acidity. However, when threshold values are exceeded the fish population starts to decline and can finally be extinct if the acidity level is high enough. The derived relationship between the level of economic activity, X, and fish stock, F, is shown in Figure 4.6. It is called a *transformation curve*. The curve starts out flat at the fish axis. The assimilative capacities result in a stable fish population up to a certain level of economic activity, and the curve ends up at the strangulation point when continuously expanding economic activity.

The welfare indicator (4.4) is represented in Figure 4.6 by contour curves, each of which has a constant welfare level. It is assumed that fish and stuff are substitutes for the people in our economy. The maximisation of welfare leads us to the point of tangency between the transformation curve between fish and stuff and a contour curve. A point on the transformation curve shows that opportunity cost of stuff in terms of foregone fish, and a point on the contour curve shows the amount of fish reduction implied by an increase of one unit of stuff for a constant level of welfare. At social optimum the opportunity cost in 'production' should be equal to the relative welfare evaluation of fish and stuff.

As is evident the actual preferences held by our society are of crucial importance as to where on the transformation curve the social optimum is located. In Figure 4.6 two sets of preferences are shown. The dashed contour curves lead to adjustment at point B and can be labelled a stuff-loving society compared with the preferences represented by the solid contour curves leading to social balance between fish and stuff at point A in a fish-loving society. How the preferences must be either for extinction or the maximal level of fish stock can easily be figured out.

Our society is seeking the solution to the following problem:

Maximise W = d(F,X)
with respect to F and X, subject to

$$SO_X = a(X) \tag{4.1}$$
$$pH = b(SO_X) \tag{4.2}$$
$$F = c(pH) \tag{4.3}$$

Insertion of (4.1)–(4.3) in the welfare function yields:

$$W = d(c(b(a(X))), X) \tag{4.5}$$

Maximising with respect to X yields the necessary first order condition:

43

$$d'_F \cdot c' \cdot b' \cdot a' \cdot + d'_x = 0 \tag{4.6}$$

or

$$\frac{d'_F}{d'_x} = - \frac{1}{c' \cdot b' \cdot a'} \tag{4.7}$$

The first term of (4.6) shows the decrease in welfare from loss of fish population caused by (starting from the right) a unit increase in consumption of stuff, leading to increased generation of sulphur oxides ($a' > 0$), which in turn leads to increased acidity in water recipients ($b' > 0$), and then leading to decreased level of fish stock ($c' > 0$), which is then evaluated by the welfare function ($d'_F > 0$). The second term shows the welfare increase of one unit increase of stuff. Social optimum is found when loss of welfare through reduced fish population just balances against increase in welfare from one unit more of stuff.

Equivalently, rearranging the condition (4.6) to the form (4.7) we have that the relative welfare evaluation on the lefthand side; the rate of substitution between fish and stuff; and the rate of transformation, via the economic-ecological systems on the righthand side, should be equal.

REFERENCES AND FURTHER READING

Førsund, F.R. (1985): 'Input-output models, national economic models and the environment' in Handbook of Natural Resources and Energy, North Holland, Amsterdam, New York, Oxford, pp. 325-41

Førsund, F.R. and S. Strøm (1974): 'Industrial structure, growth and residual flows', in J.G. Rothenberg and J.G. Heggie (eds), *The management of water quality and the environment*, Macmillan, London, pp. 21-69

Haavelmo, T. (1971): '*Forurensings problemet fra et samfunnsvitens-kaplig synspunkt*' ('The pollution problem considered from the socio-scientific point of view'), *Sosialøkonomen*, 25, 5-8

Kneese, A.V., R.U. Ayres and R.C. d'Arge (1970): *Economics and the environment. A materials balance approach*, Resources for the future, Johns Hopkins Press, Baltimore and London

Leontief, W. (1970): 'Environmental repercussions and the economic structure: an input-output approach', *The Review of Economics and Statistics*, 52, 262-71

5

Public Measures. Principles and Alternative Remedies

5.1 INTRODUCTION

In this chapter we shall outline the main principles for formulating public measures utilising the technical control possibilities for the reduction of pollutant residuals that were discussed in Chapter 4. Public control of the use of the natural environment may comprise direct control in the form of laws and order; another approach is to apply indirect measures in the form of taxes, charges, and the like.

Section 5.2 contains a general discussion of the need for public environmental protection measures. The basis of this discussion is the question of establishing control mechanisms for the allocation of scarce resources in an area where there is no reason to expect that mechanisms of this kind will arise automatically. As a background for a discussion in principle of the means used in environmental policy, there follows in section 5.3 a description of the so-called Polluter Pays Principle.

The basis for discussions of alternative means in section 5.4 is a desire to achieve socially efficient solutions. The aim is to discuss how, generally speaking, a given and desired quality for the natural environment can be achieved at the lowest cost to the community.

In section 5.5 various transitional problems in the introduction of new measures are dealt with, while section 5.6 describes various conditions that may have an important bearing on the future work of setting up public administrative units.

5.2 THE NEED FOR PUBLIC MEASURES

This need will be illustrated by discussing the consumer aspect of an economy. In a market economy the individual consumer and, by implication, the total number of consumers are not directly motivated to consider the fact that total consumption of a particular good inflicts damage on all consumers by, for example, pollution of the air in the manufacture of that good.

We will endeavour to substantiate this statement as follows. Let us assume that the ordinary consumer reasons on the following lines: 'I shall take into consideration the fact that my consumption is part of the total consumption of this good and that this total consumption injures me through air pollution.' In a case of this kind it must be presumed that the consumer is aware of the connection between the damage and total production. It is not reasonable to expect this in a market with a large number of consumers and producers.

Let us suppose that a consumer selected at random has a certain opinion or idea about this connection. If the consumer adapts his consumption of the good, he will discover that his adjustment will not have any particular bearing on total production of the good in question, and thus on the amount of air pollution. The reason for this is that his consumption comprises only a very small fraction of total consumption: the consumer is one of many consumers. This phenomenon is usually referred to as the '$1/n$' effect where n is a large number of individuals.

The individual consumer will consequently not consider it very worthwhile to reduce his consumption of the good. With a sufficiently large number of consumers it is reasonable to suppose that consumer adaptation will not result in much attention being paid to the pollutant effects of consumption. This could, in fact, occur even though every single consumer realises that consumption involves the inconveniences of pollution — for himself as well as for others.

If the consumer selected at random makes adjustments on the basis of other suppositions, the conclusion might be different. He could reduce his consumption of the good by more than the small and negligible amount mentioned above. The consumer selected at random will then take into consideration not only the fact that he himself is exposed to air pollution but also that other consumers are. However, the point is that the individual consumer has no guarantee that all consumers will undertake an extra sacrifice of this nature; for this reason only a minority, if any, will be prepared to make this act of self-denial.

All consumers will be dissatisfied with the total side-effects of the activities in which they are engaged in this economy, despite the fact that the action or behaviour of each individual, considered in isolation, is entirely rational.

There is consequently a need in this community for collective or public measures. These measures must aim to change the terms on which consumers (and producers) make their choice. This does not mean that we are suggesting a transfer of all quantitative decisions from individual consumers and producers in the economy to a central authority. Collective measures must be formulated in such a way that every consumer receives a guarantee that everyone, in making the necessary adjustments, takes sufficient account of the fact that consumption involves pollution. An appeal from a collective body to all consumers to show a certain measure of moderation in their consumption of the pollutant good could provide one example of insufficient control by the decision-making bodies. An appeal of this kind has no built-in guarantee of the kind referred to above. State intervention in the form of taxes or duties is one example of a collective measure in which market participants are induced to take into account the indirect effects of their adjustment. We shall clarify this point below.

5.3 THE POLLUTER PAYS PRINCIPLE (PPP)

As a basis for a discussion in principle of the means available in an environmental policy in the next section, we shall here give a short description of the Polluter Pays Principle (PPP).

All member countries of OECD have accepted as a principle that the polluter must pay the full cost of the environmental protection measures which the community considers it necessary to enforce. This principle is applied in direct control of discharge, orders for changes in manufacturing and production processes, and taxes imposed on pollutant discharges. However, as yet it has not been practised on the basis of a systematic evaluation of what the social costs of various forms of environmental inconveniences represent.

Many OECD countries have stated that they favour a more far-reaching interpretation of the PPP. It is maintained that the polluter, in addition to the cost of purification, should also be held financially responsible for damage caused by remaining discharge. An effluent charge would take care of this. Responsibility of this nature is rather more comprehensive than that laid down inter alia in the general rules for compensation at present embodied in most countries' legislation.

The core of the principle is that effective measures must be formulated in such a way that the costs of production of a good reflect the various forms of treatment costs and/or the damage to the environment arising from the manufacture and/or use of the good concerned. The object of this principle is to enforce economy of scarce resources, so that the social costs of achieving a given and desired environmental standard will be as little as possible. Furthermore, as a long-term measure, a principle of this nature is also intended to induce rational evaluation and choice as between a reduction in pollution and increased production in the traditional sense of the word.

It should be emphasised that the principle does not involve, for example, that air and water pollution 'must be brought to a halt'. The intention is to enforce a mechanism in which the inconveniences of various types of pollution are weighed against the cost to the community of avoiding them. The principle that the polluter should pay aims primarily at creating an economically effective use of resources in an economy, in particular the use of the waste disposal services of the natural environment. Translating this principle into terms of practical policy, whichever method is selected, might involve radical changes in production processes, changes in production and consumer patterns, and so forth.

Even though the discharger of residuals, for example, a manufacturer, will initially have to bear the burden of new and extra environmental costs, the changes in costs could have further repercussions in the economy. This could occur in several ways: one example might be effluent charges, via the market mechanism, being passed on wholly or partially to the consumers of the final product. In this way the price of goods to a certain extent would reflect the costs of the use of the waste disposal facilities of the natural environment in the manufacture of goods.

In some cases passing on the costs in this way will prove impossible. In this connection we might envisage domestic firms selling goods on an international market and obliged to accept the prices on the world market as given. Any 'environmental taxes' levied on firms of this kind would have no effect on the international commodity market. The prices of goods would . remain quite unaffected by these domestic measures.

It does not, however, follow from this that as a long-term policy the idea of imposing environmental taxes on export businesses should be rejected out of hand. Doing so would mean subsidising export businesses: they would be excused the payment of actual costs that figure in the domestic economy. In this connection it is worth mentioning that there are not many people arguing in favour of the subsidising of the use of labour in Norwegian factories, on the grounds that wage costs in Norway might be higher than in the various countries with which we compete in the world market. Sound policy in such cases might be to consider concretely the environmental problems of export businesses in every single case, so that economic consequences would be weighed against existing environmental problems.

5.4 ALTERNATIVE APPROACHES IN ENVIRONMENTAL POLICY

5.4.1 The problem

The use of indirect measures such as charges in environmental policy offers an approach that as yet has been little used in most countries. The use of charges is given a considerably more comprehensive treatment below than direct controls, which constitute more established and familiar measures. In sections 5.4 and 5.5 of this chapter a verbal pro and con discussion is pursued, while a formal treatment is offered in Chapter 6.

The following general effects of different public environmental protection measures may be of interest:

(a) effects on social efficiency of utilisation of resources, including Nature's waste disposal capacity;
(b) effects on the distribution of goods, including natural services and considerations of justice, namely, which and in what way measures affect producers and consumers.

The emphasis here will be placed on how efficient use of resources is to be achieved. The effects on distribution were briefly discussed in section 4.4 in the preceding chapter.

The market system may be said to do the main job, where efficient utilisation of available resources is concerned. From the point of view of economic theory a natural idea is to exploit and improve the market system so that it takes into account the negative environmental effects described above. In economic theories on externalities we have seen how the market system can be improved, *inter alia*, by the introduction of taxes on activities that produce negative indirect effects. A socially efficient allocation of resources can then be achieved, if the taxes are made equal to the values of the marginal damage. We shall deal with this in a number of subsequent chapters (see in particular Chapters 6 and 8).

The general starting-point in environmental policy will be to extract payment for burdening the natural environment with pollutant residuals, or, in other words, a price must be set on waste disposal services. This can either be done indirectly, by the use of charges, or directly, by direct control. Both these alternatives will, in fact, involve a certain form of pricing (cf. section 5.4.5 below).

The view that we cannot afford certain measures aimed at preventing pollution may be based on the idea that sparing Nature is a luxury — something we need not bother about. This point of view, however, will not be maintained in cases involving privately-owned resources. The reason for this attitude may be that the limited capacity of the natural environment is a collective resource which is never introduced into the market system through the medium of any private vested interests. If the waste disposal capacity proves a resource in short supply with the use of waste disposal services, then the community is faced with certain costs and collective action must be taken to incorporate these costs in the market economy.

5.4.2 Charges

The point of using taxes is to influence the decision-making units in the market system to make use of the alternative possibilities described above (see section 4.5). In using charges the information and the knowledge possessed by the individual con-

50

sumer and the individual firm on the facilities available for reducing discharges of pollutants can be utilised. This decentralisation of the decision-making process may ensure social efficacy, assuming rational action and behaviour — for example, that a given amount of goods is produced at the lowest possible cost to the firm.

From the point of view of the material balance (what comes into an activity must leave it, in terms of weight), in concrete cases it may be advisable to impose a tax on inputs in the manufacturing process — for example, use of water, addition of chemicals and such substances as mercury, chlorine, etc. If we know that inputs of this nature finally end up as residuals, imposing a tax on inputs may prove both a cheaper and a more effective form of control. Taxes will continue to play their part in encouraging the introduction of new production processes, consumption patterns, etc.

An extension of this idea is the proposal for a general tax on certain materials extracted from Nature. This tax should be imposed at the manufacturing or import stage. The tax rate is adjusted so as to equal the social marginal costs of the most deleterious deposit of these material as residuals. The charges can then be repaid, either wholly or partially, to anyone capable of demonstrating less injurious deposits. In the case of 100 per cent recycling the entire amount of the tax is refunded, but a proportion would be refunded in the case of deposit producing a certain degree of damage, and so forth. The onus of proof is in this way placed on the user of the raw material.

A method of this kind will provide socially correct signals for the utilisation of raw materials. The environmental costs of discharge will be built into the prices of raw materials, while at the same time users will be stimulated to select techniques capable of reducing the strain on the natural environment. From an administrative point of view it may prove a great advantage that the users are compelled to demonstrate their methods of deposit. In certain cases the consumption of a good may be identical with what can be discharged. One example of this is plastic bags. A duty on discharge would in such cases coincide with a manufacturing charge, provided the plastic bags are not stored and do not remain in use with the consumers. Even though there may not be concurrence between product and discharge, for practical reasons there may in many cases be a fair measure of approximation in imposing a tax on the products instead of on

51

the discharge. Examples in this category might include detergents (the pollutant discharge being phosphates and nitrates) and snow scooters (the polluting discharge being noise).

How is one to arrive at the correct level of tax? By way of general guidelines it might be worth noting that environmental costs 'originate' in the recipients. The fact that these have varying waste disposal capacities means that the size of the charge must depend on which recipient receives the residuals. For this reason a tax on, for example, wood-fibre discharge should vary according to the location of the factory; it would, for example, cost the community less to discharge a certain amount of fibre in a river with a large flow of water than would be the case in a river with a small flow of water, all other things being equal.

Deviating from recipient-based charges would cost something in the nature of reduced social efficiency. For this reason any deviation must be based on other advantages, for example, reduced administrative costs.

A practical basis for levying a charge might be to establish certain qualitative standards in the recipients, in the form of an insistence on a maximal content of various substances — for example, the oxygen content in a river should not drop below a certain level; the sulphur content in the air above a city is not to exceed a certain level on average per week; and so on. Qualitative standards or environmental standards of this kind replace what one is after in principle, namely, the individual choice of members of society between the provision of environmental services and manufactured goods and services.

For discharges to the same recipient a uniform charge per residual unit is then imposed. The point of using taxes in this connection is that one is thereby enabled to maintain these qualitative constraints at minimal social cost (see Chapter 6).

Firms have different production technologies, and the result is that they cannot reduce their discharge to the same extent. It will prove worthwhile for a firm to reduce its discharge to the point at which the cost of discharging 'the last unit', that is, marginal costs, corresponds exactly to the charge paid per unit of residuals. Since the charge rate is common to the firms involved, this means that they will reduce their discharge to the level at which they all have the same marginal costs. This in turn will mean that firms capable of low-cost purification will reduce their discharges relatively more than firms which, in the initial situation, face higher purification costs.

52

The decision on choice of technology will once again be decentralised, so that the necessary information in this case need not be centralised. The charge also constitutes a lasting economic incentive to reduce discharge of pollutant residuals. In this way firms are encouraged to invest in technological changes that will prove kinder to the environment.

In practice, it must be assumed that the controlling administrative unit is not sufficiently familiar with the natural, technical and economic circumstances to be able to decide in advance what particular charge corresponds to a particular standard. In such cases one solution is to alter the rate at certain intervals until the standard is obtained. Naturally, in such cases practical limitations will arise, in the form of restructuring costs to be borne by the firms.

Standards may be changed over a period of time in order to reflect changes in assessments of environmental services or new technology at discharge sources. By establishing the cost of maintaining standards, a better basis will be provided for a general assessment between waste disposal services and other environmental services. If, for example, maintaining a standard proves cheap, it may be sensible to raise the standard, if the discharge still results in substantial environmental damage.

An environmental policy based on public setting of standards in the recipients, and one that is to be achieved by taxing discharges, omits an important aspect of the use of the natural environment, namely, those producers or consumers who request other services apart from waste disposal services from the natural environment. On the one hand we have users of recipients for waste disposal purposes and on the other hand users of recipients for purposes of recreation and extraction. For this reason we actually have two groups of demanders, one demanding waste disposal services, the other amenity and extraction services. The supply side is to be found, so to speak, 'hidden inside the natural environment'. It is difficult in a case such as this to suggest a form of market in which sellers and buyers meet one another in the way that is common in other traditional markets. The ideal would have been a market form of this kind; however, indirect methods exist to discover, for example, what recreation consumers are willing to pay for a unit improvement in the recipients.

And yet today it would be difficult to imagine this set up for the sale of environmental services being translated into terms of

53

practical policy. Publicly established standards in the recipients are not an ideal solution; it is a practical approximation of the ideal. It contains two possible disadvantages:

standards may be fixed at the wrong level in relation to the ideal solution;
in this approximate solution only those who make use of waste disposal capacity are paying for the natural environment.

Even though the established recipient standard might coincide with the ideal solution, those who demand recreational and extractional services may demand more than is warranted by the correct solution, because no payment is made for extractional and recreational services. In such cases a lack of prices may result in actually unfounded dissatisfaction and misunderstandings.

The administrative and supervisory cost of a tax system should be built into the charges, if costs depend on the quantity of residuals. If costs are independent of the amount of residuals, there should be no addition to the charge. In such cases a total assessment must be carried out before the tax system is introduced involving weighing administrative costs and relevant current costs against the social advantages of reduced burdens on the environment.

Examples of concrete tax systems (see section 5.4.1) exist primarily for discharge of residuals in water. The reason for this is probably that in many cases the problems of measuring discharge can be overcome: the recipient is conveniently limited in the physical sense, and the number of sources as a rule is not particularly large. In the case of highly polluted and sluggish watercourses where insufficient oxygen content is often one of the main problems, it has also been possible to establish meaningful qualitative standards for the recipient. As a rule problems arise as a result of other pollutant components and effects. These conditions make it more difficult on a short-term basis to establish generally accepted qualitative standards in this country.

In the case of air pollution it may prove more difficult to introduce a tax or discharge. Pollution caused by motor vehicles presents problems with regard to the measurement of discharge: not only are the sources numerous, but they are mobile at the same time. The use of oil for heating homes and offices presents similar problems with regard to measurement and the number of sources. In both these cases it may be relevant instead to tax fuel as an input factor, for example, according to its lead and sulphur content.

In summing up we ought to emphasise the following advantages of a charge system:

charges on residuals have a direct bearing on the sources or causes of pollution;
the units possessing the technical information are motivated to introduce changes. This motivation is continuous, so that the choice of technique over a period of time will be influenced in a manner favourable to the environment;
decisions on changes in technology, etc. are carried out on a decentralised basis, so that some of the costs of a centralised collection of information will be saved.

Yet another advantage is that charges provide the controlling body with revenue. This amount might be said to represent a remuneration to natural capital, which without a tax system accrues to those responsible for pollution. Total revenue allocated to the public sector is decided at a political level. The revenue from an environmental charge would make it possible to reduce other taxes and charges collected by the public authorities.

One important point, however, is that environmental protection must not be imposed for fiscal reasons. What has been emphasised above is that for a given desired level of taxes and charges the tax revenue for a new item would necessarily mean a reduced need for tax and duty revenues from other items. There may be certain forms of discharge that are only minimally influenced by a tax, for example, household sewage. Taxes in such cases may still serve a purpose, though more from the point of view of distribution and justice. Persons responsible for pollution must pay any purification costs at central plants.

5.4.3 Subsidies

Instead of a tax being paid per residual unit discharged into a recipient, one might in principle consider firms receiving a subsidy per unit of residual that they refrain from discharging. A subsidy system of this nature may have the same short-term effect as a tax system, but may entail numerous disadvantages from the point of view of distribution and justice, and also from the supervisory point of view.

In the world of practical politics discussion has centred mainly on subsidising certain kinds of capital equipment. We shall here endeavour to evaluate whether this approach is feasible on a permanent basis. In section 5.6 we shall refer to subsidies as a means to be used during a transition period.

In section 4.5 attention was drawn to a number of technical possibilities for reducing discharge. If the public authorities compel a choice of purification equipment in cases where firms without any capital subsidies but with a tax on discharge would not have chosen purification, subsidies entail a social loss of efficiency. Given a demand for a certain reduction in discharge, subsidies may result in excessive investments in purification equipment in cases where other methods of meeting this demand would be socially more effective. In order to avoid this subsidies must go hand in hand with direct control in the form of a ceiling on the volume of investment or constraint on discharge. If a reduction in discharge is not required, investment subsidies under 100 per cent will not result in any investment in purification at all, and consequently not in any reduction in discharge either. Investment subsidies are thus not a decentralised economic instrument to the same extent as charges, but involve many of the same disadvantages as direct controls.

The technology of the firms in the long run may be compelled to adopt purification as the only alternative. Firms have no continuous encouragement to undertake any technical changes that will reduce discharge.

By imposing a constant subsidy rate per unit of reduction of discharge the contribution to the individual firm will be greater than the purification cost (see section 6.2). This will provide the individual establishment with a motive for increased production. This increase in production will occur either as a result of existing firms expanding, or by new firms being set up. In every firm discharge subsidies will result in a reduction of discharge. However, the possibility still remains that increased production, in old and new firms, may result in an *increase* of total discharge, as a result of the policy of environmental protection. One of two measures may be taken to prevent this paradoxical result; one is to introduce variable rates of subsidy, graded from zero in accordance with the reduction of discharge. A subsidy programme of this nature is difficult to carry out in practice, as it demands full information on all technical factors on the part of the environmental protection authorities. The other alternative is to introduce a system whereby potential discharge reductions, too, are granted subsidies. By potential reductions of discharge we mean discharge from firms that have not yet been set up. Everyone can threaten to set in motion pollutant production of this kind, and in principle they can also threaten to discharge unlimited amounts. This form of subsidy granted for *not* discharging residuals could easily degenerate into the absurd.

We have, in fact, two main types of subsidy: subsidies for purification equipment and subsidies for a reduction in discharge. Investment subsidies entail lesser social costs than discharge subsidies. The use of subsidies has certain distributive effects. In the first place firms that are subsidised will have better operating results than what would be represented by the social contribution. This could mean increased remuneration for owners, shareholders and employees. In the second place subsidies will probably be financed by ordinary tax revenue. This might mean that consumers who enjoy no benefits from the services from the recipient have to help to pay for an improvement in these services.

In cases where a tax on discharge was passed on, subsidies would prevent this and at the same time prevent a desired restructuring of the consumption of goods and services. Both from the point of view of efficiency and from that of fairness subsidies are therefore not a suitable means.

5.4.4 Deposits

Deposits are one way of limiting discharge of certain kinds of solid waste, that is, objects which do not change their physical shape or form when used. Examples of these are packagings of various kinds, such as bottles, tins, crates and containers. Deposit systems can also in principle be used for various sorts of consumer goods, such as cars, refrigerators, television sets, etc. A deposit is included in the purchase price and this is refunded to the purchaser when he returns the item concerned. Deposit arrangements are for this reason an encouragement to re-use and recycle. In order to prevent inflation undermining people's motivation to hand in objects at collecting points, the deposits must be index-linked. If the object has a long life, and is sold a number of times before being finally scrapped, deposit arrangements could have unintended distributive consequences (see Bohm, 1981).

5.4.5 Direct regulations

On the basis of the technical possibilities available, described in section 4.6, public authorities can also use direct controls in the form of regulations and/or bans. Every polluting unit responsible for pollution would then have to bear the cost of discharge reduction in order to achieve the set qualitative standard. Socially effective use of direct regulations places a considerable burden of information on the controlling authority. The latter, in principle, should be familiar with the technical possibilities in the case of every single source of discharge. If an order for the introduction of a particular technology is to be socially effective the administrative authorities must arrive at a technical alternative involving the lowest costs for maintaining a specified standard. A factor involving certain complications is that the various dischargers (for example, factories) have different technologies, and consequently different purification costs. As a result the public authorities will need to be in possession of a very large body of information.

The most common form of regulation is to set a particular limit for discharges and to order the introduction of specific methods of treatment. There is a danger here, however, that this may be assumed to be sufficient, without pursuing the consequences in the recipients that provide the basis for an assessment of the social costs of discharge. Establishing upper limits for discharge has the effect of making it cheaper for firms to operate at these upper limits: there is no financial encouragement present to reduce discharge by investing in new technology favourable to the environment.

Another factor in direct regulation is the administrative work in public bodies concerned with the work of dealing with applications for concessions and dispensation. This may result in too much red tape and public administrative costs.

Where an absolute ban is involved we might say that this is approximately the same as a very heavy charge. If we accept that the use of the product or substance responsible for the discharge of a residual can never compensate for the social damage, a ban of this nature would prove an effective and correct measure. This might be the case with goods for which demand is very elastic (that is, there is no difficulty in finding alternatives satisfying the same need) and/or which involve very considerable social environmental costs, even with small discharges. A ban saves administrative costs, if we except the costs of 'policing'. These, however, may well be considerable.

Mandatory alternative input factors can also prove effective. An example of this might be a mandatory transition from heavy to light fuel oil.

5.4.6 A market for discharge permits

Let us consider more closely discharges to one and the same recipient from numerous sources of discharge (for example, a discharge of organic waste from a number of factories situated round the shores of a lake). Direct regulation, if it is to be optimal, demands that the controlling authority has a detailed knowledge of every single source of discharge. Collecting information of this kind may prove very costly, and to avoid this, but without sacrificing social efficiency, the sale of discharge permits can be introduced. The controlling authority would then determine the total discharge on the basis of an assessment of recipients. This total discharge is less than the actual discharge to the recipient before controls become operative. Discharge permits are then issued. These specify how much of the pollutant the holder of the permit is allowed to discharge per time unit — for example, a year. To ensure that this system operates satisfactorily, every permit must cover a small number of pollutant units in relation to the total discharge to the recipient. Each permit specifies the same amount of permitted discharge. No one is allowed to discharge residuals without producing a discharge permit. The system is inaugurated by making permits available for sale. The point now is that a price level will emerge for discharge permits, as firms will have to bid for them (since less will be discharged than previously), and because a market will subsequently emerge among the various firms for the sale and purchase of permits.

Market equilibrium for discharge permits may be described by saying that the price of a discharge permit will be equal to the marginal costs of depositing the discharge in some other way than discharging it into the recipient (purification, recycling, etc.). If the price exceeds the marginal costs in a particular firm, it will be in the interests of that firm to purify on a larger scale than was done initially. The proportion purified will increase until the marginal costs of purification are equal to the price of the discharge permit. It will then not prove profitable to purify more, but to purchase discharge permits for discharging the remainder. If the price is lower than the marginal costs initially, the firm will increase its profits by reducing the proportion purified and instead purchasing a discharge permit. This will apply to all firms. Equilibrium is therefore achieved when the marginal costs of purification are the same in all firms. In the pollution problem proposed this precisely provides the conditions for a socially optimal adjustment. The same result could have been achieved by imposing a tax on discharge (see section 5.4.2). The advantage of negotiable discharge permits in relation to charges is that the total discharge is laid down initially, whereas in the tax solution total discharge will be determined by the reaction of firms. Adjustments in charges may prove necessary in order to arrive at the total desired level of discharge although adjustments of this kind may prove relatively expensive.

The sale of discharge permits at a later stage would be determined by new firms that are established in the area, by any expansion of existing firms and by firms cutting down on production or moving out of the area. Technical improvements in the purification processes will result in a wish for having less discharge permits. This will lead to a fall in the price of these permits.

If the authorities wish to change the total discharge level on the basis of amended assessments of environmental qualities, they can purchase and re-issue discharge permits. Purchases will render environmental standards more stringent, while fresh issues will involve a slackening of these standards. Discharge permits will thus acquire the status of securities. This, combined with the purchase and sale of permits by the public authorities, will have distributive consequences which will have to be discussed in detail before the system is made operative. The optimal quality of this system will be undermined if the market is dominated by one or a small number of firms. In such cases the condition that marginal costs in purification are equal in every firm would not be fulfilled.

An interesting possibility in this system is that outside interests, for example, a nature conservation organisation, may be allowed to buy up discharge permits. The authorities may allow this when they are uncertain as to what the desirable level of discharge should be, and in cases in which they wish to give a certain amount of latitude to individual assessments of the environmental qualities to be established. A purchase undertaken by an outside interest will mean that there will be a smaller total number of discharge permits available for firms responsible for pollution, and consequently smaller discharge in the area.

Marketable emission permits have not been applied, but a start towards such a system has been made recently in the United States in respect of air pollution. Three related measures have been developed:

(a) Emission offset

This measure is designed to make possible growth of production in an area while ensuring no further degradation of the ambient air quality. New entrants must obtain sufficient reductions in discharge of air pollutants from existing plants so that the total discharge of pollutants is equal or less than before.

(b) Bubbles

Originally this measure was introduced to allow one firm with a complex of plants or several emission sources to decide for itself where among the several sources to cut down on emissions when demanding a reduction of the *total* level. The bubble concept has expanded to include emission trades also between different plants in the same area.

(c) Emission reductions banking

This system allows sources to reduce their emission more than required, and then to 'bank' the difference for later use or sale to other firms. The idea is to give all firms incentive to develop new cleaner techniques than currently necessary.

5.5 FURTHER DETAILS OF THE CHOICE OF MEASURES

5.5.1 Introduction

The general question of the use of direct and indirect measures in environmental policy has in recent years been the subject of considerable discussion in the USA and many European countries, with the focus very much on arguments for and against the use of charges versus direct regulation.

This discussion has been carried out in various fora and on different levels. If, for example, we consider the developments in the USA, there was a great deal of resistance to the effluent charge approach at the end of the 1960s, that is, to a strategy aimed at creating decentralised economic incentives to reduce discharges of pollutant residuals. The opponents comprised not only politicians and various Nature conservation groups, but also a great many persons active in American business and industry. That the last-mentioned category should have preferred direct regulation to indirect public control with the aid of price mechanism was perhaps somewhat surprising, not least because the use of economic incentives in dealing with decentralised decision-making units is preferred in many other areas.

The use of charges as a possible strategy in environmental preservation policy has gradually been accepted in the administration. The annual report of the American Council of Environmental Quality for 1971 had the following to say on this subject: 'It is clear that because of the enforcement, efficiency, and equity problems of the regulatory approach, other means of achieving pollution abatement must also be probed.'

In England, too, this problem is given particular emphasis in public reports and findings. In the third report of the British Royal Commission on Environmental Pollution, published in September 1972, a separate report entitled 'The case for pollution charges' argues that serious consideration should be given to the introduction of charges as a realistic and practical possibility in a number of cases. Furthermore this report raises the question of why the environmental sector should be one of the few areas in which indirect public control and use of price mechanisms should *not* in principle be used for the solution of a number of problems.

The case in Norway for direct regulation versus charges was first aired in public in the government's long-term programme for 1974-7. So far the most important approaches have been individual discharge permits and regulations for purifying technology, product standards, etc. These will also apply for a number of years to come. In a government white paper (1976) relating to anti-pollution measures, however, we find the following:

> However, closer consideration will be given to the problem of whether one should not make greater use of charges on a long-term basis ... It is necessary to establish new economic mechanisms and incentives, capable of contributing to a reduction of pollution at the lowest possible cost. Charges by way of paying for environmental services and as payment for being able to utilise some of the community's common resources could be necessary measures in this connection.

In order to illustrate the practical importance of the choice between the various measures against pollution we can reproduce some calculations from a study on pollution of the Delaware delta in the Philadelphia region of the USA. The Philadelphia region is one of the most highly industrialised areas in the world. For the entire Delaware river a minimum demand for the oxygen level was introduced (3.5 ppm dissolved oxygen). The river was divided into zones: each zone has several different discharge sources. The problem is to achieve the oxygen level demanded at the smallest possible cost. The cheapest solution, assuming one possesses and utilises all the information available on the possibilities for purifying at each discharge source, was calculated. No account was taken of the administrative costs involved in implementing this detailed control solution. Three alternative measures for achieving the goal for the oxygen content of the river were then assessed. The total purification costs (that is, all costs exclusive of administrative costs and paid-up charges) were calculated.

The first measure involved demanding the same percentage reduction in discharges from the various sources (direct regulation). The second measure involved introducing an equal charge on every release of effluent in the entire area that fulfilled the oxygen level aimed at (the uniform charge solution). The third measure involved introducing a charge equal for all the discharge sources in a zone, but varying from one zone to another (zone charge solution). Table 5.1 shows the costs of these three measures as a percentage of the cheapest full information solution.

We observe that direct control with the same percentage reduction of discharge is almost twice as expensive as the cheapest solution, whereas the zone charge solution is only some 20 per cent more expensive. The reason why the uniform charge solution is somewhat more expensive than the zone charge solution (but substantially cheaper than direct control) is that a uniform charge does not take into account the fact that the location of the sources of discharge has a bearing on the effect on the oxygen content.

Below we have itemised some of the arguments most frequently used against the charge solution and in favour of direct regulation. In each case a commentary on these arguments is provided.

5.5.2 The measurement and control problem

If the basis for charges is to be the discharged amount of waste, the introduction of charges requires information on what sort of discharge firms are responsible for. If control is to be exercised it is also necessary to know how much waste they have actually discharged. Secondly, it is necessary to know how discharges influence the recipients, as ideally speaking the charge should reflect the damage that discharges inflict on other services made by recipients apart from waste disposal services. Conditions in the recipients will vary according to time and location. Topo-

Table 5.1: Costs of the three alternative purification measures, Delaware Delta, USA, 1964 (shown as a percentage of the cheapest full information solution)

Measure 1 Direct regulation	Measure 2 Uniform charge solution	Measure 3 Zone charge solution
286	171	123

Source: Kneese (1977).

graphy, weather, and other natural and local conditions will influence recipients' ability to absorb waste substances. For this reason charges must vary and must be based on objective assessments. On this basis a number of experts have rejected the charge solution.

Charges or direct regulations are means: the end is to achieve improvements in the recipients. It is not logically tenable to argue in favour of only one approach, on the ground that one does not know the effect of discharge on the state of the recipients. In introducing charges it is imperative to have some idea of the way in which discharges will affect the condition or state of recipients: frequently, this must be based on subjective evaluation, but precisely the same consideration must be taken into account in the case of direct regulation. In using direct regulation it must therefore be implicitly presupposed that it is possible to say something (subjectively evaluated) on the way in which discharges affect the condition of recipients. Precisely the same information on the interplay between discharge and condition of recipients is necessary in the charge solution. The fact that one knows nothing (or not enough) about this interplay is consequently just as much an argument against direct controls as against the charge solution. The same reasoning can be used in the argument that one must know what sort of discharge firms are responsible for when introducing a charge.

The problem of supervision is somewhat different; in the charge solution this problem will consist in ensuring that firms respect the charge rules. This means that the authorities must ensure that there is a correlation between the amount of the charge and the amount of waste discharged. This can be done in several ways: one can, for example, undertake direct observations on the spot or use charge returns and branch information (that is, what sort of discharge and what quantities are typical of the branch) combined with information on the extent to which a firm deviates technically from the branch average, or use a third (and still more indirect) method which will be discussed below.

If metering does not take place continuously, the regulating authority must work out an estimate for the total discharge in a period (a year, for example), with the aid of spot checks. This estimate will provide the basis for charges. For a factory responsible for pollution the basis for a charge will thus be stochastic. In order to prevent firms from carrying out effective purification

65

merely when spot checks are being undertaken, it is important that no advance information should be available to indicate when the inspector is about to arrive, or when he has been there. The firm has to adjust under uncertainty.

By way of example let us suppose that the discharge without purification from a factory varies over a period of a year, reaching a low level during one half of the year and a high level in the other. The inspector arrives once a year; the basis for the charge levied for the whole year will be the result of the spot check. The factory operates with subjective probabilities as to when a spot check will be undertaken, and organises its purification over the one year period in such a way that the sum of expected amount of the charge and the current costs of purification per year will be as little as possible. We shall now compare the adjustment of the factory with the solution that would have been optimal if the environmental protection authorities themselves could determine discharges over the period of one year. This optimal solution is characterised by the fact that marginal purification costs should be the same in both half-year periods, and equal to the marginal damage caused. Since we have presupposed that potentially the greater amount can be discharged in period 2, this means that it would be optimal to do so.

In this solution, the result of an adjustment on the part of the firm based on uncertainty, the marginal purification costs will be equal in the two half-year periods, if the same subjective probability for controls exists in each period. If the factory suspects that the greatest chance for control exists in the period in which unpurified discharge is greatest, it will undertake more purification than is optimal. Nor can we exclude the possibility that the factory discharges less waste during the period when the potential discharge is greatest than in the period when the potential discharge is least. Adjustment also results in marginal damage only being equal to marginal purification costs when the subjective probability is equally great for both periods.

If the factory has risk aversion, then this is tantamount to the factory in the example quoted above assuming a greater probability for control taking place in the period with the potentially greatest discharge, and it will then purify more than is optimal. Adjustment in the face of uncertainty results in the demand for efficiency not being satisfied. Control and adjustment under uncertainty will be dealt with in greater detail in Chapter 7.

It should not be forgotten that with direct regulation, too, problems of supervision will exist. We can distinguish between two forms of direct regulation:

(a) mandatory reduction of discharge, with a given number of units, or an order not to discharge more than a certain amount;
(b) order regulating the size of the discharge, combined with an order for a particular kind of purification equipment.

We shall return to the question of the efficiency of mandatory enforcement in the next secton. At the moment we shall concentrate on the problem of controls. When using the direct regulation type (a) it is left to the units themselves as in the charge solution to select the purifying technique, or, where necessary, the change in manufacturing processes, necessary to fulfil the regulatory provision.

In the case of regulation type (a), however, it is only possible to choose direct observation to check that the firms are fulfilling the regulatory provisions. *Per se*, this is also necessary in the case of regulation type (b). The special feature of (b), however, is that the provisions relating to a given purification equipment may mean that the regulatory provisions are automatically fulfilled provided the mandatory purifying equipment is installed. In some cases it may be sufficient merely to carry out one check in the factory, namely, to ensure that the mandatory equipment has been installed, but the main rule here, too, is that a check must be carried out to ensure that the purification equipment is at all times functioning as prescribed.

Let us suppose that the authorities are sufficiently well informed to know what purification equipment would be cheapest to install, given the level of discharge of waste. In such cases this particular purification equipment would have been the one the factory would have chosen for the charge solution as well. If the need for continuous supervision is superfluous in the case of control type (b), it will also be superfluous for the same reason in the case of the charge solution.

It is probable that the public authorities are not always in possession of the necessary information on what would be the cheapest purification equipment to install in the various factories. This will result in a given level of discharge being achieved at excessive cost, not only for the factory, but for the community as a whole, with a consequent waste of resources. We may therefore conclude that the charge solution does not necessarily involve greater measuring and supervision problems than direct regulation.

The objection to the charge solution has been raised that after a charge on effluent has been introduced, it is not known what the discharge will be, but this is known in the case of direct control. The problem has several aspects, however.

By 'charge and recipient standard' is meant a system in which the environment is divided into problem areas, for example, sections of a watercourse might constitute an area of this kind in which the public authorities will set a quality standard. Charges are levied on activities in the area discharging the pollutants in question. These charges can then be adjusted over a period of time, so that the quality standard is maintained. If pollution exceeds what is laid down in the quality standard, the charge may be raised. The quality standard may be violated in two ways; (a) factories may ignore the charge rules; (b) the charge is initially too low.

In the case of (a) a problem of supervision arises. Observing the quality of the recipient and then if necessary increasing the charge represents a third and indirect method of solving the inspection problem. (In such cases the increased charge has the function of a potential 'temporary fine'.) Inspection opportunities and adjustments of this kind cannot be utilised in any simple way with direct control of one of the relatively large number of activities. If this involves the charge rates having to be adjusted relatively frequently, firms may have to face problems of adjusting themselves to such frequent changes. This involves adjustment costs, which must be weighed against a purer environment.

5.5.3 The efficiency of the measures direct regulation and charges

In this section we shall deal with other aspects of the two types of measures other than problems of measurement and inspection, even though these are relevant to the problem in this section.

The basis for further discussion is that the environmental protection authorities have introduced standards for environmental qualities (see section 5.4.2). The problem then is what measures should be used so as to achieve this target as cheaply as possible. Assuming that the damage, to which discharge from a source contributes, varies from source to source, according to where the source is located, in such cases complete information is necessary on locality, manufacturing and purifying technology, and on the recipient, whether charges or direct regulation are employed. From the point of view of static efficiency it may be much of a muchness what approach is used. The advantage of a charge now lies in motivating firms to be constantly concerned to reduce their discharge still further. The special feature of direct control is that firms will find it profitable to discharge the maximum permitted. There will be no economic spur, for example, to effect changes in manufacturing processes with a view to reducing discharge beyond this amount. If direct regulation is in the form of an upper limit for discharge, then expanding firms will be motivated to attempt to improve their purification facilities.

Let us now assume that the damage caused by discharge from a number of sources depends on the total discharge, and not on discharges from the various sources. As shown in section 5.4.2 the given environmental qualities can be achieved by imposing a uniform effluent charge. One advantage of the charge solution is that the environmental protection authorities do not need any information on the manufacturing and purifying technology of the various factories.

With optimal regulation which discriminates between factories according to their purification facilities, complete information is still necessary from the various individual sources of discharge. (If complete information exists, effluent charges, too, can be used.) For this reason what must be assessed in making a choice between charges and direct regulation is the cost of trial and error to discover the correct charge as against the cost of obtaining complete information.

We shall now consider situations in which the alternative involving complete information is out of the question. A simple and administratively cheap method of control is to demand the same percentage reduction of discharge from the various sources of discharge. As shown in section 5.5.1, a solution of this kind may be substantially more expensive than the complete information solution. Of greater importance now, however, is that a charge solution, involving a uniform charge, and which is

69

based on the same amount of information (that is, information is available on recipients but no information is available on sources of discharge), may prove far cheaper than the simple regulatory solution. This may be the case even when it might have been optimal to apply individual location-determined charges. The most important reason for this conclusion is that the uniform charge results in equal marginal purification costs, and in most cases this will imply the greatest saving.

Finally in this section we shall discuss the control situation that arises when the effects of discharges in recipients are uncertain. The damaging effects, for example, of discharges of sulphur in the air may vary according to weather conditions. There may be daily as well as seasonal variations. Air inversions may produce short-term acute problems. The flow of water in rivers is subject to typical seasonal variations. For this reason the damaging effects of organic waste discharges will vary over a period of time. The point is that the waste-disposal capacity of recipients may vary over a period of a day and a year.

If the damaging effects over a year are exclusively due to extreme conditions such as atmospheric inversions and the like, direct regulation at these unexpected points of time will be the control measure involving the least cost. These short-term controls will mainly mean reducing production activity in a few places. A solution involving charges would in such cases take too long to implement. The point of charges is for them to play on the potential substitution possibilities (see section 4.6). On such short-term basis as we are here dealing with the charge, if it is to operate satisfactorily, would of necessity degenerate into a prohibitive levy. Compared to direct regulation this would prove ineffective.

If the damaging effects over the year are due to continuous load, with certain extreme and incidental peaks, a mixed control programme would involve the least cost. The charge has the effect of reducing the average level of discharge, whereas direct regulation deals with extreme and incidental situations. The use of direct regulation enforced at short notice would be particularly important if there is believed to be a considerable danger that vital threshold values involving, for example, conditions injurious to health, irreversible changes in the environment, etc. may be exceeded.

5.5.4 Will pollution remain unaffected by effluent charges?

It is maintained that some of the charges will merely result in 'legalising pollution'. Particularly large and profitable firms will merely pay the charge and continue to pollute as before, it is asserted. In answer to this it may be observed as follows:

— The introduction of effluent charges on residuals will increase costs of using certain activities within a firm. A rational firm manager must in many cases be expected to work on the basis of the following simple formula: 'Given existing unit costs associated with various production processes and the prices of products, I shall choose the processes and the pro- duction scale that enable the aims of the factory to be reached, e.g. the greatest possible profit.' We cherish a general belief that firms will be able to understand the sig- nificance of the cost factor of the environmental charge, and gradually adjust to these charges. Adjustment might, for example, take the form of altering production processes in such a way that discharge of pollutants is reduced and the charge is avoided. Another possibility might be to change the scale of production. Recycling of waste substances and the installation of purification equipment are included in the designation 'a change of production processes'.

— The possibility of immediate change of production processes may, of course, be very strictly limited in many plants. The technological structure may be so rigid that the amount of discharge can only be changed by altering the plants' scale of production. If charges result in a firm continuing to earn a current profit when full capacity is utilised, the firm will on a short-term basis continue as before. The charge will influence the assumed potential for investment in the firm, involving an assessment as to whether it would prove profit- able for the firm to change its capital equipment so that less discharge occurs. It is, of course, possible that a calculation of this kind may produce the answer that it is not profitable. The marginal damage inflicted by the pollution reflected by the charge would then not be 'big enough' to lead to changes in production equipment. The total social costs of anti-pollution measures associated with this firm will exceed the costs involved in pollution. Initiating anti-pollution measures in the firm concerned would then result in a social loss.

5.5.5 Earmarked charges

It has been maintained that if charges are to be introduced, the revenue they bring in must be 'earmarked', in other words used for environmental purposes, for example, the purchase of purification equipment and research. One argument against a proposal of this kind might be, for example, that it is perfectly possible that the need for purification equipment is not in the least related to the reduction in profit represented by the environmental charges. It should be emphasised that the primary role of environmental charges is to govern the allocation of resources in the community. Linking the level of environmental activities with the revenue from charges may result in a social waste of resources.

5.5.6 Charges and income distribution

One argument against charges that has been maintained in the general discussion on this subject is that certain goods and services will be more expensive than before, and therefore only the 'rich' will be allowed to pollute. This can be interpreted as follows: those who maintain this view believe that the distribution of income in the community is not what they would have liked after the imposition of charges. (It is possibly but not necessarily the case that they believe the distribution of income was correct before the imposition of charges.)

This, however, is not necessarily an argument against the principle of environmental charges, but an argument in favour of the public authorities not necessarily having to accept the existing distribution of income. A great many means of influencing income distribution exist; direct regulation, too, will influence the prices of goods and services, and consequently have some bearing on distribution of incomes.

Another argument against a charge solution, and one which will complicate it considerably, is that those who suffer as a result of pollution should be paid the amount of the charge by way of compensation. Even though pollution is controlled by means of charges, so that marginal damage is taken into consideration, pollution will as a rule not disappear entirely. The total damage has not been eliminated. All things being equal, those who were living in a certain locality before pollution started have been 'subject to encroachment', and should receive compensation for this, it is maintained. Once again, however, we should point out that this is a problem involving income distribution, and should be solved as such. Nor are there any grounds for automatically linking compensation with the revenue accruing from charges.

Just how much the public authorities are prepared to disburse by way of compensation is a problem of income distribution policy, and should in many cases be capable of being considered out of the context of the amount accruing in revenue from effluent charges.

Another argument against charges is that the pressure of taxes and charges on industry would increase with the introduction of environmental charges. But in the first place it should not be forgotten that direct regulation, too, will impose certain costs on industry. In addition, the charges will represent payment for any injurious effects of the discharge of residuals. Introducing the cost of using the natural environment as an economic factor necessarily involves new and extra costs for firms causing pollution, whatever approach is used, but the costs for the firm will be greater when the effluent charge is imposed than in the case of direct regulation. In principle these costs should reflect the fact that use is being made of a resource which actually is in short supply. By introducing environmental costs the community (firms and households) receives extra services in return, namely, a purer natural environment.

In the second place it should be remembered that the cost of imposing charges is not as a rule paid for by industry alone. In several cases part of the charge is passed on to the consumer. In the third place, if the public authorities discover that entrepreneurial incomes in industry are unfavourably affected by the introduction of charges, relief can, of course, be given in direct or indirect industrial taxation in order to counter this.

5.5.7 Pricing the environment an impossibility?

This argument has featured prominently in public discussion. Environmental charges entail 'pricing' the services made by the environment. Pricing the environment is impossible, according to some people. This has been interpreted in at least two ways:

(1) *'It is impossible to set a high enough price on the*

73

environment'. Spokesmen for this view are also in favour of a policy of prohibition. A total ban on discharge of a particular residual must imply that the damage caused by this discharge must be very considerable. As a general rule a total ban is, of course, an impossible rule to apply in practice. It might involve a huge number of applications for dispensation and possibly an increasing number of court cases.

In the case of certain substances and in strong concentrations it is, nevertheless, possible that the damage caused by the discharge is very considerable. Banning in such cases (discharges of toxic substances in very strong concentrations could be an example of this) is therefore the only acceptable solution.

(2) *'It is impossible in practice to price the environment'*. Without involving ourselves in a lengthy discussion on this subject, we shall merely briefly point out that direct regulation, too, implies evaluating environmental services, as mentioned above. Direct regulation of discharge will entail costs for the dischargers. Let us suppose that the permitted amount discharged from every activity influencing, for example, an amenity provided by a given recipient were to be reduced such that the amenity service provided by the recipient would be increased by one unit. If we then consider the extra total costs that this has involved for all activities reducing discharge, the amenity service of this recipient must at least be worth this amount. While direct regulation is tantamount to indirect pricing, a charge is a more direct form of pricing. The particular means applied does not depend on whether the environment in the traditional sense can be price-tagged.

5.5.8 The number of activities on which charges are imposed

Finally, we shall deal with the argument against charges which states that if there is a small number of activities or even just one polluting a given recipient, then there will be little advantage in introducing the charge solution compared with direct regulation. This may be true, at any rate in the short run.

The point is that the more activities there are contributing to the same type of pollution, the more advantages charges have, compared with direct regulation. If there is only one activity or one set of technologically similar activities responsible for

74

pollution, there is little advantage to be gained in a charge solution as compared with direct regulation. As mentioned above, cases of this kind will probably occur more frequently in small countries than in larger countries. If the total discharge is dominated by one unit, it may be socially profitable to make use of direct regulation of the one dominant unit and 'forget' the others.

One argument which still favours charges is that they might continuously result in firms keeping up to date on technical developments in their branch involving less discharge, and consequently smaller charge expenses. In the case of direct regulation firms would be able to take things a little easier, merely taking care not to discharge more than the permitted maximum. In such cases discharge of residuals would be free of charge, but not if the charge solution were in use.

5.5.9 Summing up

This chapter presents an argument in favour of a rational, permanent alternative in formulating environmental policy comprising:

the establishment of environmental ambient standards for recipients;
the imposition of effluent charges on discharges of pollutant residuals with a view to satisfying such demands for ambient standards.

We have pointed to a number of cases in which mandatory orders and direct control would prove the most sensible alternative. This might, for example, involve cases where:

the damage caused by discharge depends on random extreme situations;
the discharge is extremely injurious and the problems so acute that even small discharges cannot be accepted (toxic substances and the like);
the environmental problem resulting from pollutant discharge is so acute that, *inter alia* with a view to avoiding irreversible processes, immediate and far-reaching action is necessary;
relatively few firms exist with a fairly similar production tech-

nique, or one dominant firm polluting a given recipient. (This may be a more pressing problem in small countries such as Norway than in the large countries on the Continent and in the USA where the number and concentration of firms may, geographically speaking, be greater.)

By way of conclusion we shall repeat that the choice between direct regulation and the use of, for example, charges will naturally have to be decided in each particular case. The above arguments are intended to present the view that the charge approach is a possibility in a great many cases and should not be excluded when a choice of this nature has to be made.

5.6 INTERNATIONAL EXPERIENCE OF THE USE OF EFFLUENT CHARGES

5.6.1 Charges on discharge into water

Measures intended to influence the quality of water are of long standing. In general, the use of economic incentives is a fairly standard approach. The application of direct regulation and indirect means such as economic incentives varies from one country to another. In France and the Netherlands water policy is based mainly on the use of charges or levies on effluent. In the USA and Japan practically only direct regulation is used. A number of intermediate methods are to be found in other countries. As far as regional institutions are concerned the well-known system in the Ruhr dates back to 1913. Here, the running of purification plants etc. is financed with the help of charges levied on members. The members are municipalities and industrial firms with discharge into the waterways of the Ruhr.

In our investigation of the charge system we shall first consider how it is possible to arrive at a starting point for assessing charges — that is, the physical basis, what are the aims of the charge, and at what level it is imposed, and what institutional systems have been established.

The physical basis for charges

Water quality is difficult to define: the various user groups will

76

emphasise the different aspects of the quality of water. The use of water can be divided into three main categories: industrial, drinking water, and amenity purposes. As far as amenity activities are concerned such subjective evaluations as the water's appearance, its smell, etc. will mean a great deal. A great many substances are discharged into water. The substances whose discharge it is aimed to control are naturally the substances that have a negative effect on water quality. The choice of substances to be controlled will depend on how quality of water is defined. In the choice of definition simplicity is generally demanded, and at the same time the definition should be universally acceptable. Water quality should be measured in terms of a limited number of factors and these characteristic properties should be easily measured. One way of arriving at an accepted definition of water quality is that the parties concerned should themselves assist in defining it. This, in fact, is done in France.

The most common parameters for water quality are:

biochemical oxygen demand (BOD)
chemical oxygen demand (COD)
nitrogenous substances (N)
suspended matter (SM)
salinity
heavy metals
toxicity
temperature
acidity (pH)

Fairly complicated relationships may exist between these quality parameters. For example, temperature will affect oxygen consumption in the decomposition of organic substances; synergistic relationships of this kind have not been taken into account in practical definitions of water quality.

In France effluent charges are based on the weight of the discharge, rather than on the volume and relative contents of pollutants. The charge is based on two components, the weight of the suspended matter (SM) and the weight of oxygen required to decompose the organic material. One-third of the oxygen consumption is assumed to derive from chemical decomposition and two-thirds from biochemical decomposition. BOD is measured on a five-day basis. The formula for calculating the

77

weight on which a charge is to be levied as far as discharges are concerned will be:

$$P = \frac{COD + 2\,BOD_5}{3} + SM$$

Since 1974 salinity has also been included as a quality parameter for water. Salinity is measured by the water's conductivity. In order to obtain better control of discharges of toxic substances a parameter has been adopted also since 1974 for measuring toxic effects. The basis for measuring the toxic effects of a discharge is a test carried out on water-fleas. The basis for the toxic effect is to discover the thinning-out of a discharge required for about one-half of the water-fleas to be killed.

In the Netherlands the calculation of pollutant discharges is based on chemical oxygen consumption and discharge of nitrogen. The discharge is converted into a population equivalent. It is calculated that 180 g of this combined unit of oxygen consumption and nitrogen discharge is discharged every day *per capita*. The formula for calculating this is as follows:

$$\frac{COD + 4.57 \cdot N}{180}$$

This is a discharge in terms of grams per day. The figure 4.57 is derived from the fact that it is assumed that the nitrogen increases the consumption of oxygen by a factor of 4.57 per gram of oxygen (Kjeldahl's index). The quantity of pollutant left after biological purification is calculated by substituting biochemical oxygen consumption in the formula for chemical oxygen consumption:

$$\frac{2.5\,BOD + 4.57 \cdot N}{180}$$

In the Netherlands a substantial enlargement for the basis of imposing a charge has been implemented during the 1980s. This includes fertilisers, salts, acidity, toxic substances, bacterial content, heavy metals, radioactive substances, temperature increases in the base of coolant water, and flavouring substances. Since 1975 a charge on heavy metals and cyanides has

78

been in force in two regions in the Netherlands.

Germany, as already mentioned, has a long tradition in the use of economic incentives for improving water quality. Up to 1981, however, no common national policy was in force. The control of water quality has long been decentralised to the regional level in Germany. The association for the Ruhr has some 1,200 members; these are industrial concerns and municipalities that discharge into the river system. The basis for calculating the charge is a detailed assessment of the manufacturing processes of the various members, that is, what particular substances they discharge. The total number of employees is considered, the amount of residuals occurring in the factories, the products that the establishments manufacture, and how dangerous the residuals are. On the basis of this information the number of pollutant units each member represents is arrived at.

On a national level Germany has since 1981 had a charge based on a 'damage unit' based on suspended solids, oxidisable matter, heavy metals, mercury and cadmium, and fish toxicity. One damage unit is equivalent to each of the following discharges on a yearly basis:

1 m^3	suspended solids with at least 10 per cent organic matter
10 m^3	suspended solids with less than 10 per cent organic matter
45.45 kg	COD (chemical oxygen demand)
20 g	mercury
100 g	cadmium
1000 m^3	waste water with same toxicity (diluting factor 3)

In Canada use is made of effluent charges in municipalities comprising approximately 20 per cent of the population. A number of municipalities are planning to introduce charge systems. Charges are based on discharges plus certain threshold values. The three components used as a base are the biochemical oxygen demand, suspended matter, and oil. Threshold values have been fixed in such a way that ordinary domestic effluent is not liable to a charge.

Measuring discharge

In order to establish the basis for charges in the case of every

79

single establishment producing effluent, all discharges must in principle be measured. The volume of the discharges and the concentration of pollution are required. Measurements of discharge involve certain costs. If measurement costs are reckoned to be excessive, one possibility may be to measure only the discharge of the greatest sources of effluent, and then use, for example, discharge per produced unit as observed in the case of other, lesser, sources of pollution. A system of this kind is used in France: so-called emission factors are calculated, that is, discharge per produced unit or discharge per unit's use of certain inputs. Special measurements can be carried out by the Water Board (Agence de Bassin) or on request by a user who is not in agreement with the estimated discharge. The cost of measurement must be borne by the firm concerned if the actual pollution is lower than that estimated, and by the Water Board if the opposite is the case. With seasonal variations of discharge this must be taken into consideration when measuring; several spot checks must be taken in the course of a year. In France the charge is calculated on the basis of an average day in the month with the greatest discharge.

In the Netherlands, too, emission factors, converted into population equivalents, are used. Emission factors are expressed as population equivalents per employee, per produced unit, per unit of raw materials, per cubic metre of water used, etc. Units discharging less than 20 population units per day are charged on the basis of 3.5 population units. Discharges from the biggest discharge sources are subject to continuous measurement.

In Germany application of the charge has some complicating elements. The damage units are usually based on *expected* discharges, not actual, and official standards are upper limits.

Level of charges

Charges are levied for two principal reasons:

    an incentive to reduce discharge;
    redistribution.

By redistribution is meant that the amount collected on the basis of the charge is distributed back to the units that have been charged or are used for measures in which establishments that have paid the charge are interested. The Water Board in the Ruhr uses the money accruing from charges to finance joint measures such as the building of purification plants, construction of dams for regulating flows of water, etc.

The charge per pollution unit is calculated every year by considering the total budget and the total number of pollution units calculated for every member. The charge per pollution unit is simply the budget divided by the total number of pollution units.

This system also applies to charges levied by the regional water associations in the Netherlands. Charges imposed by the state in the Netherlands are redistributed in the form of financial aid to firms that were in existence before the Water Act came into force in 1970. In France, too, the amount of charge levied is redistributed in the form of financial aid to industry and municipalities.

It is a long-term aim, both in France and in the Netherlands, that the charge is to become increasingly a proper incentive charge, not merely intended to cover a need to finance measures for the improvement of water. On the basis of economic theory we know that the socially correct charge will be one that is equal to the marginal damage caused by discharges when the total discharge is at a desirable level. It would be pure coincidence if a charge intended to cover a given budget were to correspond exactly to this socially correct amount. Charges of this kind introduced on the basis of budgetary considerations will naturally, too, act as economic incentive. Where they go wrong is that they fail to make the proper use of the economic incentive. Distribution charges have been introduced during a period of transition to stricter environmental standards. Charges of this nature must be considered in connection with the desire to assist existing industries to adapt to new rules.

As far as the level of the charge is concerned, this varies from one part of France to another. Generally speaking, charges are not very high, accounting for approximately 0.2–0.5 per cent of industrial value added. They are regarded as too low to provide any economic incentive to persuade the individual firm to introduce extra purification measures and the like. Nevertheless, in the course of time charges have increased markedly. This is also the case in the Netherlands, implementing water pollution charges during a phased programme in the period 1970-87. The charge has risen fivefold from 1971 to 1974. In 1975 charges in the Netherlands were three to eight times higher than corresponding charges in France. In the Netherlands it is now considered that the incentive limit for charges has almost been reached. By incentive limit is meant that it is in the interests of the firm itself to reduce its discharge, for example, by installing its own purification equipment. The marginal cost of purification of this nature is lower than the charge level when no purification takes place. It will therefore pay the firm to purify right up to the point where marginal costs of purification correspond to the charge.

Institutional framework

Measures can be initiated on three levels:

(1) on the central level;
(2) on the regional level;
(3) on the local user level.

Charges have always been used in combination with direct controls. Furthermore, charges have been used at the local and regional levels, but seldom on a national level. It is only reasonable that there should be a decision-making body responsible for a naturally delimited water area: this is the set-up in France, the Netherlands and Germany. These regional bodies possess a high degree of financial efficiency in the implementation of the various water-improvement schemes are concerned. In France, direct controls are initiated by a body other than that which implements economic measures. In the Netherlands, where the same body is responsible for both kinds of measure, the state distributes support for environmental protection investments in existing industry in the various water region areas. The system in Germany means that the regional water associations impose charges on users and then transfer these charges to the regional authorities, who in turn are responsible for investing them in purification measures and the like. The existing water associations were previously independent, and did not transfer any funds to the regional level. The introduction of systems of charges also has a political aspect; in France, for example, local mayors have opposed the introduction of charges because this is unpopular among the electorate. In designing a system of charges a considerable emphasis has therefore been placed on ensuring that it is politically acceptable. Allocation to the parties involved of the charges collected is regarded as a great advantage where political acceptability is concerned.

Concluding remarks

Effluent charges are most generally used to control the quality of the water, but even in this respect there are relatively few countries with nationwide comprehensive systems of charges, namely, France, the Netherlands and Germany. The systems of charges are still relatively new, and the charge levels are generally too low. The greatest effect is achieved by the redistribution of charge funds. The level of these charges, however, is still too low to act as a correct economic incentive. In the Netherlands, however, a certain degree of incentive effect has now been recorded.

The basis for charges is calculated according to well-tested rules, and does not appear to involve major problems. The system, too, does not appear to be expensive to run. In France, for example, the administrative costs account for about 5 per cent of the total costs to the regional water board. Integrated control at the level of a natural water area appears to be the most promising basis for controlling water quality. It is also a good basis for the introduction of effluent charges.

As far as the political success of introducing charges is concerned, three factors play an important role, namely, decentralisation, financial independence and government participation.

5.6.2 Charges levied on discharges to air

Little information is available on the international level with regard to the experience gained from charges levied on discharge to air. There are, in fact, only two countries for the time being where what might be called a system of charges levied on discharge to air is in use. These countries are Norway and the Netherlands. One reason why discharges are less commonly used for this purpose may be that air space as such has no natural regional boundaries, as is the case with watercourses. The pollution of the air over a given area from a given discharge source will vary with wind and weather. For this reason there is a far more complicated relationship between discharge of various substances and the qualities of the air than would be the

83

case with water. It is worth mentioning, however, that in America the country has been divided into 247 air quality control regions.

In countries where effluent charges have been levied on discharges to air, or where proposals have been made for introducing these, sulphur has been substance on which the authorities have concentrated on a discharge. In principle, the basis for the effluent charge is the amount of sulphur that a discharge source discharges per time-unit. There are two types of discharge sources for discharges to air, namely, mobile and stationary sources. So far it has only been suggested that stationary sources should be subject to charges. Mobile sources, that is, motor vehicles, would be regulated by means of direct technological restrictions, rather than an effluent charge.

The pollution effects of sulphur discharges can be divided into two types:

(1) short-term effects, caused by local concentrations and resulting in corrosion damage to buildings and the like, and increased incidence of respiratory diseases;
(2) the other detrimental effect of sulphur discharge is the more long-term acidisation of water and soil. Owing to the spread of sulphur through the air this more long-term effect is of a rather more national character as compared with other local discharge problems.

Reducing total discharges

With a view to reducing the more long-term acidisation of the environmental measures to influence total sulphur discharges provide a natural solution. The basis for a charge, as already mentioned, is in principle the amount of sulphur discharged by the discharge sources. However, registering this may well involve measurement problems. An alternative method might be to consider the sulphur content in the inputs used, for example, the sulphur content in oil. There are reasons to believe that direct measuring methods would prove cheaper and more convenient in the long run.

In the Netherlands the various types of fuel determine the basis for levying a charge. The charge is not constant per unit of sulphur in the type of fuel concerned, but is a fixed charge per weight or volume unit of the fuel type. The charge levied on the various types of fuel is adjusted according to its average sulphur content. The fuel types involved are petrol, paraffin, light oil, diesel oil, heavy fuel oil, gas and coal.

In Norway, too, the sulphur content in oil provides a basis for the sulphur charge. But in Norway the charge is proportional to the weight of sulphur in the oil, if we ignore the fixed charge of one øre per litre of oil. The Norwegian charge system was substantially improved in September 1976: firms that were in a position to prove that their discharge of sulphur was lower than the sulphur content in the oil would warrant received a refund of the charge proceeds in relation to the reduction in sulphur discharge. A reduction of this kind could be achieved by, for example, purification, or by binding sulphur in the product, as is done in the manufacture of cement. The amount refunded is of the order of 1.5 million kroner annually. With a refund arrangement of this nature the mineral oil charge in principle acts as an effluent charge on sulphur caused by the use of oil as an input. The Norwegian system has aroused international interest. In Sweden it is proposed to introduce a system similar to the Norwegian one.

Determining the level of the effluent charge

The general principle for determining an effluent charge, as already mentioned, is that the charge is to correspond to the marginal damage caused by the discharge, at the level of total discharge which is considered socially correct. As far as discharge to air is concerned it may be difficult to apply principles of this kind, even when the long-term effect on the natural ecology is concerned, because the relationships are not known accurately enough. A better operational basis would be to decide the total discharge level considered desirable in the country per annum. The charge can then be adjusted so that this total discharge level is reached. In order to know what charge should be fixed, it is essential to find out how the discharge sources, that is, the firms, react to the imposition of a charge. Firms' possibilities of reducing sulphur discharge will generally involve

transition to fuel-types with a lower sulphur content;
reduction of discharge by purifying stack gases;
reduction of the sulphur content by purifying the sulphur at the fuel production stage;

changes in production processes involving less input of sulphurous factors;
reduction of total fuel consumption.

If the charge is to produce a lower discharge of sulphur, it must be fixed at a level so that at least one of these possibilities will be utilised. Economic reason presupposes that at any rate on a long-term basis firms will make adjustments so that their marginal costs, in reducing sulphur discharge, will balance the effluent charge.

Generally speaking, charges are low both in Norway and in the Netherlands. As far as a change over to lighter oils is concerned, for example, the charge in Norway on heavy oils is so low that these would still be economically worthwhile in firms that make use of heavy oil in their production processes. For the time being purification costs are higher than the differences in price between heavy and light oils. In the long term this might well be changed, in particular with the introduction of thermal power stations, which can wash the stack gas in sea water. Desulphurisation of oil is today more expensive than purchasing light oils. In the long run, with increased demand for light oils, it is reasonable to believe that the differences in price between heavy and light oils will be determined by desulphurisation costs. This presupposes that the demand for light oils with a low sulphur content is greater than natural sources of light oils. In addition to the low levels at which effluent charges are fixed in Norway and the Netherlands there remain the effects on total fuel consumption and conversions in manufacturing processes. This latter aspect is particularly important in the long run.

Reduction of local concentrations of sulphur

Effluent charges have not been established in any country with the special objective of reducing local concentrations of sulphur. In principle we might imagine a regional differentiation of the sulphur charge. This would then have to be combined with direct controls, in order to regulate such factors as the stack height discharge sources would choose. Direct regulations are also needed in influencing local concentrations, because a charge can hardly be changed as rapidly as might be necessary in order to deal with changes caused by varying wind and weather conditions, for example, occasional air inversion during the winter months.

86

Concluding remarks

We have so far concentrated on sulphur as an air discharge variable. It would be relevant to introduce several other types, too. The interplay between various types of discharge, such as, for example, the formation of photochemical smog, is a special problem.

Another problem is the measuring of discharge. An alternative to continuous measurement would be, for example, in the case of discharge to water to use certain discharge coefficients, namely, discharge of various substances in relation to, for example, total production, total employment, and the like. It would then be up to firms who are not in agreement on these calculations to prove that they discharge smaller amounts. With regard to discharge levels in the countries where a charge is in use, Norway and the Netherlands, it might in general be said that they have been far too low to be able to have any bearing on the possibilities available to firms for the reduction of discharge. No basis of experience on the way in which effluent charges can work will be available before these charges reach a reasonable level of incentive.

In particular with regard to controlling local concentrations of air pollution it would be necessary to combine the use of charges with direct regulations, owing to the many haphazard circumstances which have a bearing on, and determine, the actual degree of pollution.

5.6.3 Charges levied on noise

In recent years there has been increasing international interest in considering the possibilities of using economic incentives to reduce noise levels. In September 1975 a law was introduced in the Netherlands making it possible to impose a charge on noise from sources defined as noisy on the basis of standards for decibel levels indoors and out. The basis for this charge is to be the duration of the noise and particular characteristics of it. Only the central government may impose noise charges.

In actual use noise charges are limited to aircraft. At the Charles de Gaulle airport in Paris a charge was imposed on landing, depending on the noise level of the aircraft. This is also the case in Japan.

As far as the Netherlands is concerned, the planned programme for combating noise might result in charges on, for example, cars accounting for as much as 3 per cent of the price of the car and about 11 per cent of the price of a motor cycle. The amount raised by these charges is to be used to improve old buildings by means of increased noise insulation and the building of anti-noise screens along highways etc.

5.7 TRANSITIONAL PROBLEMS

The introduction of new measures will naturally present the parties concerned with problems of adjustment and conversion. In principle the basis of special transitional arrangements only exists if private conversion falls short of what is socially desirable and capable of being carried out.

In using charges, for example, any transitional arrangement envisaged must not prevent the general encouragement to invest in the development of new and environmentally friendly technology, etc. Transitional arrangements should, where possible, avoid retaining a structure incompatible with the change in the composition of goods generally aimed at.

Such measures may involve firms having to carry out a number of new investments in the form of purification plant, new production capital, etc. Existing capital equipment must be replaced more rapidly than would have been the case without such measures. This will involve firms in an increased need for loans. Should the money market not be operating effectively it may be necessary to assist firms with loan facilities.

The subsidising of investment in purification plant operates as a transitional arrangement in a number of countries, such as Sweden. However, the same objection to subsidies as a transitional arrangement may be raised as is the case with the permanent arrangement dealt with in section 5.4.3. In order to arrive at an effective utilisation of the community's resources, firms should be responsible for all social costs. Generally speaking it should be possible for these to be covered if the firm concerned is to justify its existence.

In this connection it is worth mentioning that in Italy a combination of charges and subsidies has been proposed as a transitional arrangement for firms discharging organic substances in water. These subsidies are paid out as a once-for-all grant for reinvestment, whereas charges on residuals are at the same time levied currently. In determining the amount of the subsidy the total amount paid out during the transitional period corresponds to the total amount collected in charges during the same period. In this way firms that are the first to invest in purification plant enjoy a bonus at the expense of firms that are slow to make this investment. In a situation involving a high capacity utilisation in the economy the total pressure exerted by an increased investment of this nature must also be considered. A gradual application of measures of this nature, stretching over a certain period of time, might well be an advantage.

In introducing new measures some firms may be forced to suspend or radically reduce production. This, *inter alia*, will mean that employees will be compelled to find other places of employment or even to be declared redundant. For this reason an overall assessment of such disadvantages, as opposed to the increased environmental qualities, must first be undertaken, in order to see whether a shutdown is really socially desirable, and in the second place an appraisal must be made to decide whether a basis exists for special supplementary measures aimed at reducing transition problems.

From the point of view of efficiency general supplementary measures are preferable, for example, loans. If the special transitional problems involve a lack of alternative employment for the labour force, the most effective solution would be to subsidise the labour force.

Areas with a weak economic basis are the ones that are particularly liable to lack alternative possibilities for employment. Here, considerations of regional policy must be weighed against environmental considerations. However, it should be pointed out in some cases that when charges are adjusted to the capacity of the individual recipients to receive residuals, in a number of cases the recipients in remote districts will possess a greater waste-disposal capacity than recipients surrounding towns and built-up areas. In this case one is thinking in particular of regions with lakes as recipients for effluent and regions with good air circulation suitable for localising air-polluting types of industry. The demand for pure air and pure water, however, may be greater in such areas on account, *inter alia*, of recreational and nature conservation interests, and water intake downstream.

There is, of course, no general clash between environmental conservation policy and the goal of full employment. However, environmental protection policy involves a change in the direction of expansion of the economy, whereby a number of employees will have to change their place of work. For this reason an effective labour market policy during a transitional period is an important precondition for environmental protection measures.

5.8 INSTITUTIONS

So far we have spoken somewhat loosely of 'the public authorities', 'the environmental protection authorities', and so forth, to describe the administrative unit capable of taking action. Just what administrative set-up would best serve the efficiency of the community is a question that should be examined in greater detail. We shall here merely mention various circumstances that may have an important bearing on this question.

The socially negative effects of discharge are bound up with conditions in the various recipients and the consumer interests therein. For this reason effective measures should be based on a wide measure of local information. This suggests a decentralised administrative set-up, for example, in the form of bodies responsible for managing the use of the various recipients concerned, estuaries, air space over an urban region, various stretches of a river, endangered lakes, and the like. Recipient bodies of this nature have been set up for watercourses in England, France and Germany. As mentioned in Chapter 3, certain possibilities also exist for influencing the self-purification capacity of recipients. Collective measures of this kind might be natural task for a recipient body. Other collective measures include the central purification plant. It would also be possible to establish plants for the recycling of useful substances in residuals.

The material balance described in Chapter 3 emphasises, however, the fact that residuals do not disappear, even though they are purified. The recipients' earth, water and air must be considered in an overall assessment if, for example, the situation is to be avoided in which the solution of an air pollution problem, by means of scrubbing of flue gases, were to produce a water pollution problem, or if the incineration of solid waste were to produce an air pollution problem.

Recipients may cut across existing regional administrative units. For this reason the establishment of decentralised recipient bodies may create special problems, which space does not permit us to discuss in greater detail.

5.9 THE ANATOMY OF ECONOMIC 'SOLUTIONS' TO POLLUTION PROBLEMS

5.9.1 Introduction

The economists' concept of a 'solution' to pollution problems generally means that the opportunity costs of using the environment as a sink for residuals, i.e. making use of waste disposal services, is at the level found to be the social optimum. A very simple model for deriving this optimum, and an analysis of how to achieve this by using a charge on residuals, are presented below.

5.9.2 The optimal level of pollution

The setting of our model may be at the macro level, for society at large, or at the micro level of one polluting economic activity, e.g. a firm. The following variables are employed:

Z = discharge of pollutants (in physical units)
$\bar{Z}$ = discharge of pollutants with no purification activity
C = costs of purifying pollutants (in money units)
D = damage of pollutants discharged to the environment (in money units).

The relationships between these variables are:

$$C = C(Z), \; C' < 0, \; C'' > 0 \tag{5.1}$$

The amount of pollutants, Z, is the amount actually reaching the environment after purification. Purification is here meant in a general sense of utilising all possibilities of reducing primary and secondary discharges mentioned in section 4.6, such as reducing level of production, substitution of input factors, installation of purification equipment and so on. It is assumed that the cost function C traces the least expensive way of

91

reducing discharge of secondary pollutants. By construction we have

$$C(\bar{Z}) = 0,$$

and over the range zero to $\bar{Z}$ of Z the larger the amount of pollutants discharged the smaller the purification costs, $C' < 0$. It may be reasonable to assume that this negative marginal purification cost increases with Z, i.e. the larger amount discharged the less the costs decrease.

The damage in the environment is related to the amount discharged through the damage function (5.2). This function measures the opportunity cost in monetary terms of using the waste disposal services. It is, of course, a heroic assumption that such opportunity costs can be measured in money units and related to discharges in a unique way. Marginal damage, D', is assumed to be positive and increasing, $D'' > 0$, when discharge to the environment increases.

$$D = D(Z), D' > 0, D'' > 0. \tag{5.2}$$

The two relations are put together in Figure (5.1). The development of the graph of the purification function is left open as very high levels of purification are achieved. For some type of pollutants complete purification is almost physically impossible and/or extremely expensive. The maximal amount discharged to the environment is $\bar{Z}$.

Figure 5.1: The optimal level of pollution

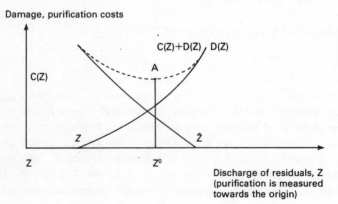

Damage, purification costs

Discharge of residuals, Z
(purification is measured
towards the origin)

The graph of the damage function exhibits the same properties as the physical relationship between discharges of pollutants and negative effects on services of the environment shown in Figure (4.2). For small amounts of discharges up to Z the environment has a free waste disposal capacity, and then damage is generated.

A natural objective function for our society is to minimise the sum of purification and damage costs. Increased purification decreases damages in the environment, but resources used up in purification activities have opportunity costs in our society measured just by the cost of purification. The optimal amount of pollutant is found by solving the following problem:

$$\underset{Z}{\text{Minimise}} \left\{ C(Z) + D(Z) \right\} \tag{5.3}$$

The necessary first order condition for an inner solution on the range $(\underline{Z}, \bar{Z})$ is

$$C'(Z) + D'(Z) = 0 \tag{5.4}$$

The absolute value of the marginal purification cost should equal the marginal damage. If we consider purifying more than the amount $(\bar{Z} - Z^0)$ satisfying condition (5.4), then the marginal cost of purifying (absolute value) would be greater than the marginal damage, and if we consider purifying less then marginal purification costs (absolute value) would be less than marginal social damage. The amount Z^0 represents the socially optimal level of pollutants. The opportunity cost of environmental degradation is balanced at the margin with the opportunity cost of purification. The sum of the cost components is at its minimum.

The sufficient second order condition for a minimum is:

$$C'' + D'' > 0. \tag{5.5}$$

By our assumptions both terms are positive.

5.9.3 Implementation of the optimal solution

It is now assumed that the relationship (5.1) holds for just one

93

decision-making unit called a firm. If a charge, t, has to be paid per unit of pollutants discharged to the environment the firm has to decide whether to discharge and pay the charge or purify. We assume that the firm acts rationally and seeks to minimise the sum of charges and purification costs:

$$\underset{Z}{\text{Minimise}} \{t \cdot Z + C(Z)\} \tag{5.6}$$

The necessary first order condition for a solution on the range $(0, \check{Z})$ is

$$t + C'(Z) = 0 \tag{5.7}$$

The firm minimises charges and purification costs when the charge per unit of pollutants equals the marginal purification cost (in absolute value). Purifying more implies that marginal purification costs (absolute value) exceeds the charge, and purifying less leads to marginal purification cost (absolute value) being smaller than the charge. Therefore, a cost minimising firm is assumed to obey (5.7).

The sufficient second order condition for a minimum is

$$C''(Z) > 0 \tag{5.8}$$

We have assumed increasing marginal purification costs (i.e. decreasing absolute value of marginal purification costs when increasing discharge of pollutants).

Figure 5.2: Implementation of socially optimal level of pollution

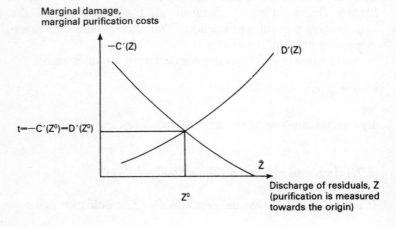

94

Since the firm adapts to satisfy (5.7) the socially optimal level of pollution is implemented when the charge is set equal to the marginal damage at the amount of pollutants Z^0.

The situation is portrayed in Figure 5.2 where marginal purification and damage curves are shown.

A more extensive treatment of charges as instruments, corner solutions etc. is offered in the next chapter.

REFERENCES AND FURTHER READING

Baumol, W.J. and Oates, W.E. (March 1971): 'The use of standards and prices for the protection of the environment', *The Swedish Journal of Economics*, vol. 73, no. 1, 42-54

Bohm, P. (1981): *Deposit-refund systems*, The Johns Hopkins University Press, Baltimore, London

Bohm, P. and C.S. Russell (1985): *Comparative analysis of alternative policy instruments*, in A.V. Kneese (ed.) *Handbook of natural resource and energy economics*, North Holland, Amsterdam, 395-460

Freeman III, A.M. and Haveman, R.H. (July 1972): 'Residual charges for pollution control: a policy evaluation', *Science*, vol. 177

Førsund, F.R. (1973): 'Externalities, environmental pollution and allocation in space: a general equilibrium approach', *Regional and Urban Economics*, 3 (1), 3-32

Kneese, A.V. and Bower, B.T. (1968): *Managing water quality: economics, technology (?), institutions*, Johns Hopkins Press, Baltimore

Kneese, A.V. (1971): 'Environmental pollution: economics and policy', *The American Economic Association*, papers and proceedings, 61, 153-66

Kneese, A.V. (1977): *Economics and the environment*, New York: Penguin

Kneese, A.V. and Schultz, C.L. (1975): *Pollution, prices and public policy*, Brookings Institution, Washington DC

Muraro, G. (1972): 'Anti-pollution policy and cost allocation: the issues in practice' in *Problems of environmental economics*, OECD, Paris, 41-57

Mäler, K.B. (1974): *Environmental economics: a theoretical inquiry*, Johns Hopkins Press, Baltimore

OECD (1978): *President's council on environmental quality. Environmental quality 1971*, Government Printing Office, Washington DC

OECD (1985): *Environment and economics*, Paris, 255-73

Scott, A.D. (1972): 'The economics of international transmissions of pollution' in *Problems of environmental economics*, OECD, Paris

Solow, R. (1972): 'The economist's approach to pollution and its control', *Social Science*, vol. 47, no. 1

Lord Zuckerman and Beckerman, W. (September 1972): 'The case for pollution charges', *Minority Report*, Royal Commission on Environmental Pollution, *Third Report*

6

Static Pollution Models

6.1 INTRODUCTION

We shall now more formally and with the use of models deal in greater detail with the concept of external effects, and consider what use may be derived from this concept in the analysis of current pollution problems.

The definition we shall use is as follows: external effects are those that other consumers and/or firms exercise on a firm's technical production environment and/or a consumer's level of utility. Moreover, the party involved cannot control the level of external effects without the use of resources.

The fundamental starting point has been to establish that if the consumers/firms in control of the factors producing external effects do not consider the benefit/damage their actions may inflect on others, the result could well be a misallocation of resources in the community. By misallocation of resources is meant that, for the same given amount of resources, we could have had a higher level of production of, for example, a particular good, whereas the production of other goods has not declined.

External effects can be divided into four types, depending on who creates such effects, and who receive them. We may have external effects as

(1) between consumers
(2) from consumers to producers
(3) from producers to consumers
(4) among producers.

In the first three sections partial pollution models will be analysed. In the last section, however, a general static equilibrium model will be used in the analysis of pollution problems.

6.2 PARTIAL POLLUTION MODEL

We shall now consider a simple partial pollution model. By partial is meant that prices are kept constant, unaffected by the variables in the model. We can assume a situation in which a number of factories or plants are discharging a residual that affects the condition of a recipient. The plant is described in terms of three relations: a cost function which tells us how much it costs to produce the product in the cheapest possible way; a discharge function that acts as a link between production and the discharge of residual; and a purification cost function, which tells us how much it costs to purify. The environment is described in terms of an environmental indicator function, which provides a link between discharges and conditions in the environment

$$c_{xj} = c_{xj}(x_j) \qquad (6.1)$$
$$z_j = z_j(x_j) \qquad (6.2)$$
$$c_{pj} = c_{pj}(z_j - d_j) \qquad (6.3)$$
$$R = r(d_1, \ldots, d_n, k) \qquad (6.4)$$

c_{xj} = the cost in pounds sterling of producing x_j in the cheapest possible manner;

x_j = the production in physical units of the plant (factory) number j;

z_j = generation of residual in physical units from plant number j;

c_{pj} = purification costs in pounds sterling of purifying the amount $z_j - d_j$ of residuals generated in plant number j;

d_j = residual discharge in physical units to the recipient from plant number j;

R = environmental indicator in physical units for the condition of the recipient;

k = purification, tidying-up activities, etc. in physical units in the recipient.

97

All magnitudes apart from R are flow magnitudes, that is, amounts per time-unit. R is a stock magnitude. For simplicity's sake we shall furthermore assume that all physical magnitudes are one-dimensional, but no problem would be involved in extending this to cover multi-dimensional concepts. As far as relations are concerned, we have in particular assumed that the purification process is additive, that is, it can be separated from the rest of production. As an argument in the purification cost function we have primary discharge minus what is actually discharged into the recipient, the residual discharge. We assume that the amount purified is deposited in a way that does not affect the recipient we are considering or the environmental indicator included in (6.4). We might also imagine a situation in which external costs associated with discharges of what we are purifying can be included in the purification cost function. Moreover, we disregard the transitorial aspects connected with (6.4) and assume that the underlying dynamic relations are stable.

With regard to the form of relations, we shall assume that the cost function for the product has continuous derivatives of the first and second order, and that the average cost curve is U-shaped. The discharge function (6.2) is assumed to have a positive derivative of the first order. The same is assumed of purification cost function (6.3). In addition it will often be natural in cases of this kind to presume that the derivative of the second order is positive, that is, that the purification cost function is convex. The environmental indicator function is specified in such a way that the various discharges may have different effects on the environmental indicator, even though the same substance is discharged. This may be due to different localisation of the plants.

The problems to be discussed are as follows:

(A) What should the purification levels be like for the individual factories in order to maintain a given environmental quality in the cheapest possible manner, when production is given for each factory?

(B) How is production to be adapted in relation to purification with a given environmental quality?

(C) The assessment of measures in the actual recipient versus individual purification.

(D) Means.

(E) Strategy for a recipient body.

(A) Minimisation of individual purification costs

The objective function is the sum of the individual purification costs. We assume here that production is given. This implies that primary discharge from each factory is given. This is indicated by a bar over the z. The restriction in the problem is that we are to maintain a given environmental quality, $\bar{R}$. Formally, the problem can be described as follows:

$$\underset{d_1, \ldots, d_n}{\text{minimise}} \quad \sum_{j=1}^{n} c_{pj}(\bar{z}_j - d_j)$$

subject to

(i) $\bar{R} = r(d_1, \ldots, d_n, k)$
(ii) $0 \leq d_j \leq \bar{z}_j \quad j = 1, \ldots, n$

The Lagrange function for the problem is (constraint (ii) is dealt with separately below):

$$L(d_1, \ldots, d_n) = \sum_{j=1}^{n} c_{pj}(\bar{z}_j - d_j)$$
$$+ \lambda_1 \{ \bar{R} - r(d_1, \ldots, d_n), k) \} \tag{6.5}$$

The variables in the problem are residual discharges from each factory. Differentiation of the Lagrange function for residual discharge from factory number j gives the following necessary conditions as an answer to problem (A).

$$\frac{\partial L}{\partial d_j} = - c'_{pj} - \lambda_1 \frac{\partial r}{\partial d_j} \overset{\geq}{\underset{\leq}{=}} 0 \text{ as } 0 \overset{d_j=0}{\underset{d_j=\bar{z}_j}{\leq d_j \leq \bar{z}_j}} \quad j = 1, \ldots, n \tag{6.6}$$

where

$$c'_{pj} = \partial c_{pj} / \partial(z_j - d_j).$$

99

Note that $\bar{z}_j - d_j$ is equal to the amount of purified residuals. Purification costs are increasing with the amount of purified residuals, $c'_{pj} > 0$. The marginal impact of residual discharges on the environmental indicator is negative, $\partial r / \partial d_j < 0$.

With increased purification of discharges from factory number j we manage to reduce residual discharge d_j. The increase in costs of a marginal reduction of residual discharge from number j is c'_{pj}. The improvement in the quality of the environment due to this marginal reduction of the residual discharge is $\partial r / \partial d_j$. The marginal damage involved in increasing residual discharge from unit number j, calculated positively, is equal to $- \partial r / \partial d_j$.

For the d_j-value that solves problem (A) the following applies:

$$\frac{\partial \left(\sum_{j=1}^{n} c_{pj}(\bar{z}-d_j) \right)}{\partial \bar{R}} = \lambda_1. \tag{6.7}$$

The shadow price λ_1 expresses changes in the total purification costs for a change in the constraint; if we increase the standard for environmental qualities, $\bar{R}$ increases. The total costs of purification

$$\sum_{j=1}^{n} c_{pj}$$

will then also increase. This means that the shadow price λ_1 is positive. The function of the shadow price is to convert the constraint to the same unit of measurement as the objective function, that is, pounds. We shall call the expression

$$\lambda_1 \left(-\frac{\partial r}{\partial d_j} \right) \tag{6.8}$$

for the marginal damage function associated with the discharge from factory number j.

Let us consider the case in which equality applies in relation (6.6). We then get:

$$c'_{pj} = \lambda_1 \left(-\frac{\partial r}{\partial d_j} \right), \quad j = 1,..,n \tag{6.9}$$

The interpretation of (6.9) is that the marginal purification cost is to be made equal to the marginal improvement in the environmental indicator, positively calculated, caused by the reduction of residual discharges through purification, and converted into pounds via the 'shadow price', λ_1, on the environmental indicator constraint.

From (6.9) we see that if the effects on the environmental indicator are independent of which factory is responsible for the discharge, that is, if $\partial r/\partial d_j$ is equal for all j, then all marginal purification costs will be equal for the factories that purify. This result is of great importance to suggestions for the imposition of charges in environmental policy. We shall return to this later. If the discharges of the factories have different marginal effects on the environmental indicator, on account of, for example, different locations, then marginal purification costs must be different. But we can note that the marginal cost per unit increase in the environmental indicator caused by purification

$$\frac{c'_{pj}}{-\dfrac{\partial r}{\partial d_j}} \tag{6.10}$$

must always be equal for all factories.

The situation involving an interior solution, that is, where equality applies to the right-hand side in relation (6.9), is shown graphically in Figure 6.1. Marginal purification costs are here shown as a function of a residual discharge. The amount that has to be purified is thus measured from point $d_j = \bar{z}_j$, and moves to the left in the direction of origin. The marginal damage function corresponds to relation (6.8) with partial variation in residual discharge from unit number j. The marginal functions are drawn in accordance with the fact that the purification cost function is convex in the amount purified, and that the environmental indicator function is concave in residual discharge. Figure 6.1 shows the typical pollution situation generally presented in textbooks. But it might be of interest, too, to look at so-called 'corner' solutions, that is, when we do not have an internal solution of the kind shown in the figure. In Figure 6.2 the entire course of the marginal purification cost curve lies below the marginal damage curve. That means that the right-hand side in equation (6.6) cannot be fulfilled with equality. The right-hand side will be greater than zero. This implies that the residual discharge should be put at its lower limit, that is, zero.

Figure 6.1: Optimal purification determined by equating marginal purification costs and marginal damage

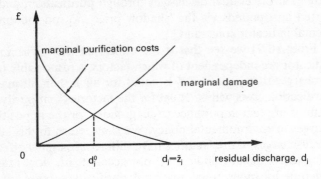

This gives 100 per cent purification. The special feature of Figure 6.2 is that marginal damage starts at a value in excess of zero. This is not as special as might at first be believed. We must remember that we are considering here a partial variation in residual discharge from factory number j. We assume that discharges from other factories are set at the level they achieve in the optimal solution. From the purely technical point of view we are talking of how the partial derivative of the environmental indicator function develops when the residual discharge from factory number j tends towards zero.

In Figures 6.3a and 6.3b we see that the marginal purification costs for the permitted values of discharge are greater than the marginal damage. In Figure 6.3a the marginal damage is zero for an amount of discharge greater than the maximum discharge

Figure 6.2: 100 per cent purification

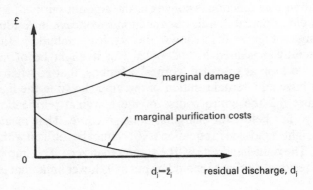

Figure 6.3: (a) No purification. (b) No purification

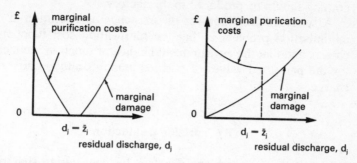

from factory number j. In Figure 6.3b marginal purification costs at maximum discharge $d_j = \bar{z}_j$ are positive and greater than the marginal damage. In these situations the right-hand side in relation (6.6) will be less than zero. This implies that nothing should be purified.

It is also possible that no solution exists for our minimisation problem. In Figure 6.4 we see that the marginal damage increases above all limits for a value of discharge lower than the maximum $\bar{z}$, while at the same time marginal purification costs rise above all limits for a value of discharge which is greater than the value for the discharge where marginal damage rises above all limits. As an example we might take a toxic substance which it is impossible to purify 100 per cent, while at the same time marginal damage rises very steeply for relatively small amounts of discharge. We must here remember that a restriction we have imposed is that production in every single factory is to be constant. If we relaxed this assumption, we would in a

Figure 6.4: No optimal solution exists

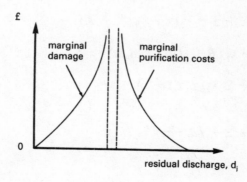

situation such as that shown in Figure 6.4 arrive at the solution that nothing should be produced in this factory.

A sufficient condition for the existence of a solution in our minimisation problem is that the purification cost function is convex and that the environmental indicator function is concave for the permitted values of purified amounts and residual discharge.

(B) Variable production

We shall now extend our problem by considering that production, too, can be adjusted. As an objective function for the problem we shall use maximum net profits in factories when purification costs are taken into account. The problem can be formulated as follows:

$$\text{maximise} \atop x_1,\ldots,x_n,\, d_1,\ldots,d_n \atop z_1,\ldots,z_n \qquad \sum_{j=1}^{n} \left[p_j x_j - c_{xj}(x_j) - c_{pj}(z_j - d_j) \right]$$

subject to

$$r(d_1,\ldots,d_n,\, k) \geq \bar{R}$$
$$z_j(x_j) = z_j, \quad j=1,\ldots,n$$
$$d_j \leq z_j, \quad j=1,\ldots,n$$
$$x_j,\, z_j,\, d_j \geq 0,\, k=0.$$

The relevant Lagrange function is

$$L = \sum_{j=1}^{n} \{ p_j x_j - c_{xj}(x_j) - c_{pj}(z_j - d_j) \} \qquad (6.11)$$
$$- \lambda_1 \{ \bar{R} - r(d_1,\ldots,d_n,k) \}$$
$$- \sum_{j=1}^{n} \lambda_{2j}(z_j(x_j) - z_j)$$
$$- \sum_{j=1}^{n} \lambda_{3j}(d_j - z_j)$$

The necessary first order conditions are

$$\frac{\partial L}{\partial x_j} = p_j - c'_{xj} - \lambda_{2j} z'_j \leqq 0 \quad (=0 \text{ or } x_j=0) \tag{6.12}$$

$$\frac{\partial L}{\partial z_j} = - c'_{pj} + \lambda_{2j} + \lambda_{3j} \leqq 0 \quad (=0 \text{ or } z_j=0) \tag{6.13}$$

$$\frac{\partial L}{\partial d_j} = c'_{pj} + \lambda_1 \frac{\partial r}{\partial d_j} - \lambda_{3j} \leqq 0 \quad (=0 \text{ or } d_j=0) \tag{6.14}$$

The shadow price is 0 when the corresponding constraints are not effective. In our problem λ_{3j} will be 0 when the situation is such that residual discharge is less than the primary residual. Let us consider a factory that is to operate $x_j > 0$ and where $d_j < z_j$. The internal solution can then be characterised with the help of the following relations:

$$\lambda_1 \left(-\frac{\partial r}{\partial d_j}\right) = c'_{pj} = \frac{p_j - c'_{xj}}{z'_j} \tag{6.15}$$

If we consider the first equation we observe that the marginal costs for a discharge into the environment (calculated positively), the marginal environmental damage, should be equal to the marginal purification cost. The costs of discharging into the environment are calculated at the shadow price, λ_1. This now measures the increase in the total net surplus for all factories, if we make a marginal adjustment in environmental restrictions, that is, if we increase $\bar{R}$. The second equation in (6.15) tells us that marginal purification costs, that is, pounds per unit of purified residual, should be equal to marginal profits before marginal purification costs have been covered, per unit of the primary residual created by a marginal increase in production. Thus the first equation determines the relationship between residual discharge in the environment and the damage inflicted on the environment. The other equation provides the balance between what should be produced and the primary generation of residuals. Expressed in other words, an adjustment should be made so that *total* marginal costs are equal to the product price, where the total marginal costs now consist of the

105

marginal cost of the goods c'_{xj} and the marginal purification cost $c'_{pj} \cdot z'_j$. The marginal purification costs consist of two parts: the marginal generation of primary residuals z'_j and the corresponding marginal costs c'_{pj}. The reasoning in the interpretation of the second equation in (6.15) is that residual discharge is constant, so that the marginal increase in the primary residual is purified in its entirety.

As in the previous problem, here, too, we can achieve 100 per cent purification. This will apply if the following holds good:

$$c'_{pj} < \lambda_1 \left(- \frac{\partial r}{\partial d_j}\right) \quad \text{when } d_j = 0 \tag{6.16}$$

This situation corresponds to Figure 6.2. In the problem set out in (B) we may pose the question whether production levels and thus generation of primary residuals are affected. Since the factory is being operated, the relation (6.12), and consequently (6.13), apply with equality. On the basis of relation (6.12) this furthermore implies that the share of marginal purification costs of the total marginal costs is positive. If we consider a situation without purification this can be regarded as though a shift had taken place in the marginal cost function for the good. A shift of this nature implies that production is less than it would be without purification costs. Even though we have 100 per cent purification there are consequently economic reasons for reducing both production and the generation of primary residual.

The situation involving zero purification can arise in the two ways shown in Figures 6.3a and 6.3b. In problem (B) z_j is no longer a given magnitude. The question then is what new element is introduced if the maximum discharge z_j can be varied. In the case shown in Figure 6.3a the marginal damage is $\lambda_1(-\partial r/\partial d_j) = 0$ for all $d_j \leqq z_j$. It is then once again obvious that environmental considerations have no effect on the adjustment of the factory. This applies even though factories adapt their maximal discharge. Discharge never exceeds the capacity of free waste disposal services.

In Figure 6.3b we have the following situation:

$$0 < \lambda_1 \left(- \frac{\partial r}{\partial d_j}\right) < \lim_{d_j \to z_j} c'_{pj} (z_j - d_j) \tag{6.17}$$

If we put together (6.13) and (6.14) when $d_j = z_j > 0$ we get

$$\lambda_{2j} = c'_{pj} - \lambda_{3j} = \lambda_1 \left(-\frac{\partial r}{\partial d_j} \right) > 0 \tag{6.18}$$

The shadow price λ_{3j} constitutes the difference between marginal purification costs and marginal damage. The shadow price λ_{2j} of generating primary residuals is positive. This means, if we consider the relation (6.12), that production and generation of primary residuals will be reduced in relation to the solution we would have if we made an adjustment of such a kind that the price corresponds to marginal costs in the production of the goods as in Figure 6.3a. Even when there is zero purification the environmental costs will nevertheless result in production and consequently discharge being reduced.

(C) Recipient measures

We shall now consider a situation in which something can be done with the recipient. The recipient's waste disposal capacity can be influenced by various measures. Examples of such recipient measures could, in the case of water, be the addition of calcium in order to counteract acidisation, increasing the flow of water in rivers in periods of low water in order to counteract the reduction in the oxygen content of the water; another example might be the injection of air in order to ensure increased circulation and increased oxygen intake, or more concrete physical changes might be envisaged in conditions in the recipient, for example, the blasting of thresholds in fjords that constitute a barrier to the circulation of water. Where the recipient is the ground itself, measures might be simple ones such as tidying up and gathering garbage, empty bottles, etc. With regard to the recipient air it has actually been suggested that a giant fan could be installed in order to increase circulation in Los Angeles. Note that all the recipient measures here mentioned are in terms of current input. We are assuming that capital investments have been translated into terms of annual costs. The problem can then be expressed as follows:

$$\text{Maximise} \atop {x_1, \ldots, x_n, z_1, \ldots, z_n, \atop d_1, \ldots, d_n, k}} \quad \sum_{j=1}^{n} \{ p_j x_j - c_{xj}(x_j) - c_{pj}(z_j - d_j) \} - p_k k$$

107

subject to

(i) $r(d_1, \ldots, d_n, k) \geqq \bar{R}$
(ii) $z_j(x_j) = z_j, \quad j=1, \ldots, n$
(iii) $d_j \leqq z_j, \quad j=1, \ldots, n$
(iv) $x_j, z_j, d_j, k \geqq 0$

The difference from the problem in the preceding section is that we now have an additional factor in the objective function, namely, costs of our recipient measures, $p_k k$, and in addition we must now, of course, adjust k itself. We shall not rewrite the Lagrange function. The first order condition for adjusting recipient measures will be:

$$\frac{\partial L}{\partial k} = -p_k + \lambda_1 \frac{\partial r}{\partial k} \leqq 0, \quad (=0 \text{ or } k=0) \tag{6.19}$$

Recipient measures should be undertaken to an extent that the marginal effect of these measures, evaluated by the shadow price λ_1, corresponds to the unit price of the input, $\partial r/\partial k$ being positive. If equality cannot be achieved, collective measures should not be applied. With regard to evaluating recipient measures and restrictions on the discharge of the individual factory, we can put together (6.14) (for $d_j < z_j$) and (6.19) and we then get:

$$\lambda = \frac{c'_{pj}}{\left(-\dfrac{\partial r}{\partial d_j}\right)} = \frac{p_k}{\dfrac{\partial r}{\partial k}} \tag{6.20}$$

As we can see, all magnitudes are denominated in terms of pounds sterling per environmental indicator unit. Relation (6.20) tells us that the costs of a marginal environmental improvement, with increased purification, must be equal to the costs per unit of environmental improvement by recipient measures. In the environmental indicator function we have substitution between discharges and recipient measures. This now implies, in relation to the situation in the previous sections, that production, primary discharge, and residual discharges can be increased somewhat, as the resultant decrease in the quality of

STATIC POLLUTION MODELS

the environment can be compensated by means of recipient measures. With regard to the environmental indicator function, we assume that this has its maximal value when the environment is untouched.

$$r(0, \ldots, 0, 0,) = R^{\max} \tag{6.21}$$

When there is no discharge, there is no need either for collective recipient measures.

If we have the situation in Figure 6.3b, the result we may get is that no discharge is to be purified in the various factories, but that collective recipient measures should be applied. With such measures, that is, an increase in k, the marginal damage function in Figure 6.3b will be shifted to the right. This will partially contribute to an increase in the shadow price λ_{3j} in relations (6.13) and (6.14). This furthermore contributes to the production and generation of primary residuals being somewhat greater, as will be seen from relation (6.12), than in the previous section. Relation (6.18) shows that λ_{2j} will be less if λ_{3j} increases. This partial reasoning presupposes that cost functions are convex and the environmental indicator function concave.

In the case of 100 per cent purification, as shown in Figure 6.2, the introduction of recipient measures may mean that we get a certain discharge. This will be because marginal damage function is then shifted to the right. The new function can now cut the marginal purification cost function. This will depend on what we assume to happen with marginal damage when the

Figure 6.5: Optimal purification before and after recipient measures

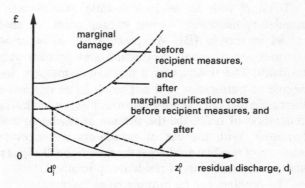

109

discharge is zero. It would be reasonable to suppose that marginal damage with zero discharge is reduced, the greater the measures we apply. With the increase in the generation of primary residuals the marginal purification cost function will also move outward. The situation can now be described as in Figure 6.5.

If we have a situation with some purification, but where the marginal purification cost is positive at zero purification, we can, by applying collective recipient measures, achieve a situation in which there is to be no purification in the various factories. This will be the case if the price of the aggregate recipient input good is lower than the marginal purification costs at zero purification in the various factories.

(D) Implementing the optimal solutions

Above we have arrived at the optimal solutions. The question is now how these solutions are to be put into practice or implemented. A distinction may be made between direct means such as mandatory measures, bans, and the like, and indirect means which aim to exploit the economic adjustment mechanisms of firms, for example, by introducing charges. The degree of centralisation governing these decisions has to be considered. In principle mandatory measures/bans are capable of an exact adjustment, whereas indirect measures, such as a charge, leave it to the individual firms to adjust their production and discharge. The simple model we are working with cannot do full justice to the discussion that has been going on with regard to the use of mandatory measures/bans as opposed to charges or other indirect means.

To start with let us look a little more closely at direct mandatory measures. As our starting point, we shall take the model in section (B). We presuppose, in other words, that factories are to adjust production and primary generation of residuals. The reason why a mandatory measure for a certain degree of purification will not result in an optimal solution in this case is that since factories are to adjust production and primary generation of residuals, the volume of discharge will be undetermined. With the use of mandatory measures for optimal amounts of residual discharges the factories will adjust production and primary amounts of residuals optimally.

The problem can be formulated as follows:

Maximise $\quad [p_j x_j - c_{xj}(x_j) - c_{pj}(z_j - d_j)]$
$x_1, \ldots, x_n, \; z_1, \ldots, z_n,$
$d_1 \ldots, d_n$

subject to

$z_j(x_j) = z_j$
$d_j \leqq \bar{d_j}$
$x_j, \; z_j, \; d_j \geqq 0$

The same problem applies to all factories; $\bar{d_j}$ in this case is the optimal solution for residual discharges that we found in section (B). The Lagrange function for the problem is

$$L = p_j x_j - x_{xj}(x_j) - c_{pj}(z_j - d_j)$$
$$- \gamma_{1j}(z_j(x_j) - z_j) \qquad (6.22)$$
$$- \gamma_{2j}(d_j - \bar{d_j})$$

The first order conditions are:

$$\frac{\partial L}{\partial x_j} = p_j - c'_{xj} - \gamma_{1j} z'_j \leqq 0 \quad (=0 \text{ or } x_j = 0) \qquad (6.23)$$

$$\frac{\partial L}{\partial z_j} = -c'_{pj} + \gamma_{1j} \leqq 0 \quad (=0 \text{ or } z_j = 0) \qquad (6.24)$$

$$\frac{\partial L}{\partial d_j} = c'_{pj} - \gamma_{2j} \leqq 0 \quad (=0 \text{ or } d_j = 0) \qquad (6.25)$$

We shall now concentrate on an interior solution. The corner solutions follow analogously from what we discussed in section (B). From (6.24) and (6.25) we see that in optimum the shadow price for primary generation of residuals will be equal to the shadow price for residual discharge. In other words, $\gamma_{1j} = \gamma_{2j}$. If the shadow price γ_{2j} is equal to the marginal damage in the optimal solution, $-\lambda_1 \partial r / \partial d_j$, the solution for the primary amount of residuals in this implementation model must be equal to what we found in section (B), because the marginal purification cost will then be the same in the two situations. Since the constraint on residual discharge here is effective, we know

111

that the firm chooses the solution for residual discharge which is equal to the optimal solution in section (B). When the solution for z_j is the same, it follows of necessity from the relation between the primary generation of residuals and production that the solution for production must be the same. The question then is whether in the Implementation Model D we shall have the same solution for x_j and z_j as in Model B merely by introducing a demand for a maximal residual discharge corresponding to the solution for residual discharge in Model B. The answer is yes. To prove this we try whether a higher z_j and therefore a higher x_j could be a solution to our problem (remember that z_j and x_j are moving in the same direction). From the second equation in (6.15) written in the form $c'_{pj}z'_j = p_j - c'_{xj}$ it follows that the left-hand side is greater, then c'_{xj} must be smaller. Since the c'_{xj} function is strictly convex, this implies that x_j must be less. This is a contradiction. If we try with a lower z we shall get a similar type of contradiction.

A central environmental authority can, of course, also make direct mandatory orders governing the size of production together with that of residual discharge. But we see here that this is actually not necessary; it is sufficient to make orders for residual discharge. The firm itself will then adjust production and generation of primary residuals optimally. We shall now see how the effluent charge will work.

(E) Effluent charges

The problem (E) can be formulated as follows:

Maximise $\quad [p_j x_j - c_{xj}(x_j) - c_{pj}(z_j - d_j) - t_j d_j]$
$x_1, \ldots, x_n, z_1, \ldots, z_n,$
$d_1, \ldots, d_n$

subject to

$$z_j(x_j) = z_j$$
$$x_j, z_j, d_j \geqq 0$$

We assume that the firm accepts the charge t_j as permanent. The Lagrange function for the problem is:

$$L = p_j x_j - c_{xj}(x_j) - c_{pj}(z_j - d_j) - t_j d_j \tag{6.26}$$
$$- \gamma_1 (z_j(x_j) - z_j)$$

The necessary first order conditions are:

$$\frac{\partial L}{\partial x_j} = p_j - c'_{xj} - \gamma_{1j} z'_j \leqq 0 \quad (=0 \text{ or } x_j = 0) \tag{6.27}$$

$$\frac{\partial L}{\partial z_j} = -c'_{pj} + \gamma_{1j} \leqq 0 \quad (=0 \text{ or } z_j = 0) \tag{6.28}$$

$$\frac{\partial L}{\partial d_j} = c'_{pj} - t_j \leqq 0 \quad (=0 \text{ or } t_j = 0) \tag{6.29}$$

We shall concentrate on an interior solution, that is, that relations (6.27)–(6.29) apply with equality. We shall then see from (6.29) and (6.28) that the charge t_j is equal to the shadow price of primary generation of residual γ_{1j}. If we compare this with the model in section (C) and look at the relation (6.14) there, we shall find that the charge in our problem will now realise the optimal solution from section (C) if the level at which the charge is set satisfies:

$$t_j = \gamma_1 \left(-\frac{\partial r}{\partial d_j} \right) \tag{6.30}$$

The solution is illustrated in Figure 6.6.

Figure 6.6 corresponds to Figure 6.1. If the charge is somewhat higher than t_j^0, we shall see that it will be worth the firm's while to discharge less than the optimal amount. Correspondingly, if the charge is set at a rate lower than t^0_j, it will be worth the firm's while to discharge more than the optimal amount. The firm now takes on the purification costs equal to the area C, and must pay the amount of the charge, equal to areas A+B. The firm pays more in tax than the damage represents. The excess amount paid is represented by area A. We can consider this as payment for exploiting a resource in short supply,

113

Figure 6.6: Optimal effluent charge

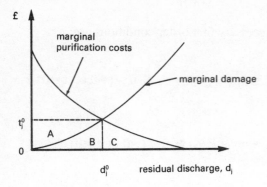

namely, our recipient. The amount, corresponding to area A, may be called a scarcity rent, which now accrues to the authority imposing the charge.

From (6.30) we can see that the charge will be the same for firms if the marginal effect on the environmental indicator function is the same no matter from which factory the discharge comes. This applies if it is the *sum* of discharges that has a bearing on the environmental indicator R and not the individual discharges.

In order to discover the optimal solution in selection (B) we assumed that we had all the information on production functions and cost functions. The solution of the problem gave us the optimal charge to impose on the firms in this section. We can note that just as much information is required in order to fix optimal charges as to fix optimal mandatory measures relating to discharge. The exception is when it is possible to know the sum of the profit functions for the various firms without proceeding by way of adding up the separate profit functions. If this should be possible, and this appears comparatively improbable, it is possible to find the optimal charge if we know the sum function and if we know the environmental indicator function. But this information is not sufficient to fix an optimal mandatory measure for every single firm. We would then have to know every single profit function. With complete information in the static case there is therefore no special point in introducing charges instead of direct mandatory measures corresponding to the optimal solution for every firm. But in cases involving incomplete information charges may have an advantage. We are

Figure 6.7: Cost effectiveness with a uniform effluent charge

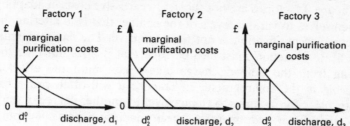

now assuming that it is the sum of discharges that influences our environmental indicator, that is, where the discharge comes from is of no consequence. Let us assume that we do not know the environmental indicator function either. If we then introduce a charge chosen arbitrarily we know nevertheless that the solution we now obtain is achieved in the cheapest possible way. The situation for three firms with different types of marginal purification cost functions is shown in Figure 6.7.

The environmental quality R that is produced is achieved in a least cost way in the sense that the sum of profit-*exclusive* duty payments is the greatest possible. This will be seen by comparing conditions (6.12)–(6.14) with conditions (6.27)–(6.29). The charge solution will minimise total purification costs. Figure 6.7 clearly illustrates the inefficacy of insisting that every factory is to reduce its discharge by the same percentage from a situation where there were no restrictions on discharge. We see that factory 1, in an optimal solution, is obliged to purify essentially more than factory 2, which in turn has to purify proportionately more than factory 3. This is determined by the shape of the marginal purification cost curve. In Figure 6.7 the broken lines show the discharge for the same percentage reduction for all three factories. Total discharge is to be the same as in the optimal solution. We note that for factory 2 the solution is precisely the same, whereas factory 1 purifies less than in the optimal solution, and factory 3 purifies more. We can see from the form of the cost function for factories 1 and 3 that it is immediately clear that this solution is more expensive in the form of higher total purification costs than the optimal solution. How much more expensive depends on how different the two marginal purification cost curves are.

We can now imagine the environmental authority varying the charge in order to arrive at the desired level for environmental

115

indicator function. This is only capable of being carried out in practice if factories are not subject to excessive costs in adapting themselves to various levels of the charge, that is, if altering the degree of purification is not too expensive. The environmental indicator function must also be such that it does not take too long from the time discharge takes place until the effect is visible in the environment. In practice it will often prove too expensive for factories to change over to another degree of purification, once investment has been made in a particular purification plant. One might after all imagine these iterations being carried out only on paper, as far as factories are concerned. They are faced with payment of a charge and are asked to state how much they would discharge, given this charge. The authorities can then evaluate the total discharge reported in the questionnaires. Even if the environmental indicator function were unknown, the aim is that total discharge is to be reduced. The advantage of the procedure outlined above is that the authorities are not required to acquaint themselves with the various purification technologies, but all the time it is known that the total discharge, occurring at every stage of charge levied, minimises total purification costs. A question arising in this connection would be whether firms could be expected to supply correct answers every time returns were called for. The firms would know, however, that in the last resort a final regulation would be drawn up, involving some form of control, so that no immediate advantage would accrue from deliberately supplying erroneous information.

One qualification must be made in the use of the charge solution. Turning back to Figure 6.6, we note that an amount corresponding to Area A, which is part of the total charge A+B, will not correspond to any damage caused by discharge. It is then possible for a situation to arise that will compel the factory to close down, because Area A exceeds the net profit. This, however, would not be socially optimal. In the event of insufficient information on the marginal damage function, it would be difficult to decide whether it is socially optimal or not for firms registering a private economic loss to be closed down. As a rule of thumb it might now be said that for firms insisting that they will have to close down if the charge is introduced, an investigation can be made to discover whether a refund of *one-half* of the charge amount A+B would be sufficient to ensure that the factory involved continues operating. Taking one-half of the charge amount means that in this case we have approached the marginal damage function with a straight line running from the origin to the point of intersection between the marginal damage function and marginal purification costs.

We can note that when using direct regulation the authorities will also have to find an optimal distribution of discharge among the various factories. Optimal distribution can be found by minimising total purification costs with the total discharge level as constraint. We can now get the reverse situation from the charge solution, namely, that factories that continue operating under direct regulation should close down on the basis of a socio-economic appraisal. From Figure 6.6 we have seen that with the use of direct regulation factories are not charged with amount B, which is an estimate of the total damage caused by residual discharge. In order to arrive at the optimal solution on this problem of direct regulation it is, in fact, not sufficient merely to minimise total purification costs subject to the discharge constraint; we must also investigate whether every single factory can afford the damage amount B. If the environmental indicator function is not known, the same approximation can be carried out as above, because in solving the problem of minimising purification costs a shadow price for restricting discharge will occur. This shadow price will be identical with the charge, given that the discharge solution is the same.

We have shown that, in order to arrive at the optimal discharge for every single factory, the use of a charge solution will demand just as much information as direct regulation. If we know the sum of the profit functions, without having to find out the profit function of each individual factory, the charge solution will require less information. In the event of insufficient information, the charge solution gives a least cost solution. This cannot be achieved by means of direct regulation.

The charge solution and direct regulation operate differently as far as the factories' profit is concerned. In the regulation solution factories pay merely for purification, whereas they do not pay for the damage caused by residual discharge. In the charge solution factories also pay for the damage caused by residual discharge, but with a normal shape of purification cost curves and marginal damage curves they pay for more than the damage. The implication is that in the case of direct regulation some factories which should be closed down are able to operate, whereas in the charge solution it may happen that some

117

factories which ought to continue operating will close down. The statement that a given reduction of discharge is achieved in the cheapest possible manner by using the charge solution presupposes that no factory is closed down merely because of the extra burden this charge involves represented by Area A in Figure 6.6. In the case of factories threatened with closure, the current profit can be corrected by applying the rule of thumb that half the amount of the charge should be refunded.

Over a period of time direct regulation and charges may have different effects. Charges provide a constant spur to improving purification technology, and no activity is required on the part of the environmental protection authorities themselves. Let us imagine that over a period of time an improvement takes place in the purification technology, as a result of which there is a change in the marginal purification cost curve as shown in Figure 6.8 by the broken purification cost curve. The optimal initial situation is that the amount d_j^3 is discharged. After the firm has recorded an improvement in the purification cost curve, we see that, with the same charge t_j, it will be optimal for the firm to discharge the amount d_j^1. With direct regulation in the initial situation for residual discharge d_j^3 the introduction of new technology will not influence the discharge if controls are maintained. d_j^3 will still be discharged. If the marginal damage

Figure 6.8: Effluent charge with a change in purification technology

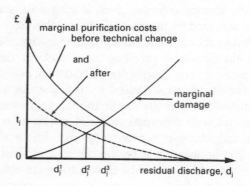

function remains constant, we can see from Figure 6.8 that it is now not optimal for the firm to discharge the amount d_j^1. The charge solution produces an automatic reduction of discharges by means of improved purification technology, but this reduction may be too great in relation to what is to be optimal. In order to fix a correct charge now, the authorities will constantly have to keep in touch with current purification technology. This means that the authorities must at all times be in possession of the same information, in order to impose optimal direct regulation, as for the imposition of optimal charges.

(F) The adjustment of the recipient authority

We shall now widen the scope of our problem by considering the adjustment of the recipient improving input. How are we to initiate or implement the optimal solution in section (C)? As an institutional frame we might imagine that there is an environmental authority or a recipient body responsible for the recipient we are considering. With direct regulation a deficit will occur, corresponding to the expense incurred in the recipient improvement. For this reason it might often be desirable that this expense should be covered in some way or other; and it would be particularly appropriate for this expense to be covered by polluters. The imposition of a charge has often been introduced in order to cover expenses of this nature.

Let us consider the implementation of the optimal solution in section (C) with the introduction of charges. By assuming that firms maximise their net profit with the charge as given, we arrive at relation (6.30), which shows the optimal charges. Another method of organising this would be to imagine that there were two types of environmental protection authorities, a superior authority and a local authority responsible for the local recipient. If we now assume that the superior authority has examined the calculation in section (C) and arrived at the optimal charges by relation (6.30), we can furthermore assume that the superior authority orders the local recipient authority to apply these charges. At the same time the superior authority orders the local recipient authority to maximise the profit, consisting of the sum of charges minus the expenses involved in the recipient activity. The problem can then be set out as follows:

Maximise $\sum\limits_{j=1}^{n} t_j d_j - p_k k$
$d_1, \ldots, d_n, k$

subject to

(i) $r(d_1, \ldots, d_n, k) \geqq \bar{R}$
(ii) $d_j, k \geqq 0$

The relevant Lagrange function is

$$L = \sum_{j=1}^{n} t_j d_j - p_k k \qquad (6.31)$$
$$- \mu_1 \{\bar{R} - r(d_1, \ldots, d_n, k)\}$$

The necessary first order conditions are:

$$\frac{\partial L}{\partial d_j} = t_j + \mu_1 \frac{\partial r}{\partial d_j} \leqq 0 \quad (=0 \text{ or } d_j=0) \qquad (6.32)$$

$$\frac{\partial L}{\partial k} = -p_k + \mu_1 \frac{\partial r}{\partial k} \leqq 0 \quad (=0 \text{ or } k=0) \qquad (6.33)$$

If we consider relations (6.14) and (6.29), and compare these with relation (6.32), the result will be that the optimal solution for discharge will be realised when the charge is fixed in accordance with (6.30). Furthermore, relation (6.33) will correspond to relation (6.19). We get the correct recipient input when the shadow price in the problem here is the same as the shadow price in the problem in section (C). As a supplementary precondition we must introduce the possibility of the recipient body operating at a loss.

It is not possible to allow the local recipient body itself to determine the charges. Maximising the profit when charges are variable would produce unduly small discharges in relation to the optimal solution. Adjustment for the recipient body would then be analogous with the adjustment of a monopolist.

We can note that both the superior authority and the local recipient authority must be familiar with the environmental indicator function, whereas only the superior authority needs to know the factories' production functions and purification cost functions. The local authority does not need to know these provided it is informed of the optimal charges by the superior authority. In the event of insufficient information on factories' product and cost functions, but with information on the environmental indicator function, an iteration of the kind described above might occur, namely, with the environmental protection authorities experimenting with various levels of the charge. As mentioned above, it is reasonable to suppose that this would be a desk operation. Factory conversions would prove too costly for an unduly frequent change of the charge.

The local environment authority decides the amount of residuals the recipient is to receive. When charges are optimal initially, these amounts will correspond exactly to what factories wish to discharge. But with insufficient information and experimenting with charges, we must continue until we achieve a balance between what factories wish to discharge and what the local environmental authority wishes to receive.

Let us now consider the financial results of our local recipient body. In the optimal solution we get the following result:

$$b = \sum_{j=1}^{n} t_j^0 \, d_j^0 - p_k k^0$$

$$= - \mu_1 \sum_{j=1}^{n} \frac{\partial r}{\partial d_j} d_j^0 - \mu_1 \frac{\partial r}{\partial k} k^0$$

$$= - \mu_1 \left(\sum_{j=1}^{n} \frac{\partial r}{\partial d_j} d_j^0 + \frac{\partial r}{\partial k} k^0 \right)$$

$$= - \mu_1 \varepsilon \, \bar{R} \tag{6.34}$$

We have here inserted relations (6.32) and (6.33). We could also have inserted relation (6.30) for optimal taxes and relation (6.19) for the connection between the price of recipient measures and the marginal utility of recipient measures. In the optimal solution μ_1 is equal to λ_1. We have assumed that these relations apply with equality, that is, we have an interior solution and in particular the constraint for the environmental indicator function will be binding, so that we have inserted $\bar{R}$ in the last expression in relation (6.34). The magnitude ε is called the *passus coefficient*. It measures percentage changes in the output when all inputs increase by 1 per cent. The product here consists of environmental indicator units, and the inputs are discharges and recipient input k. The passus coefficient measures the percentage change in the environmental indicator for 1 per cent change in all discharges and the recipient input. We see from (6.34) that the budgetary situation is directly linked with the sign of the passus coefficient.

$$\text{Budget} \begin{cases} \text{surplus:} & \varepsilon < 0 \\ \text{balance:} & \varepsilon = 0 \\ \text{deficit:} & \varepsilon > 0 \end{cases}$$

The fact that the passus coefficient, for example, is negative means that the same percentage increase in discharge as in collective measures produces a deterioration in the environment. The environment indicator registers a decrease. This may very often be a usual situation. We shall give examples of all three types of budgetary situation.

Example 1. Cleaning up. The recipient authority's books are to show a profit. We shall now assume that it is only the sum of discharge that is of any significance to the environmental indicator, that is, it is of no consequence what each factory discharges. Any input made in the recipient is of such a kind that we are able to remove some of the discharge in the actual recipient. If, for example, we suppose that the discharge consists of paper waste in a natural park, the recipient input consists of tidying up this garbage. The unit measuring the recipient input can in this case consist of hours of work spent in tidying up. We furthermore assume that every input of the unit in the recipient removes a unit of the waste. The remaining pollutants in the recipient, which have a bearing on our environmental indicator, are the total discharge minus $a \cdot k$. The environmental indicator function will be a monotonic transformation of the remaining amount of pollutants:

$$r(\cdot) = r(ak - \sum_{j=1}^{n} d_j) \tag{6.35}$$

The partial derivatives of this function are

$$\frac{\partial r}{\partial d_j} = -r', \quad \frac{\partial r}{\partial k} = r'a, \quad r' > 0 \tag{6.36}$$

We presupposed that r' was to be positive.
It is impossible to tidy up more than is discharged.

$$\sum_{j=1}^{n} d_j \geq ak \tag{6.37}$$

The insertion of (6.36) in the passus equation gives us the following equation:

$$\varepsilon \cdot \overline{R} = \sum_{j=1}^{n} \frac{\partial r}{\partial d_j} d_j + \frac{\partial r}{\partial k} k$$

$$= -r'(\sum_{j=1}^{n} d_j - ak) \leq 0 \tag{6.38}$$

It will be seen from equation (6.37) that the passus coefficient in this case must be less than or equal to 0. It will be 0 if everything discharged is tidied up. The conclusion here is that normally the budget for the local recipient body will operate with a surplus, and will balance with complete collective purification or tidying up.

Example 2. Dilution. The recipient authority's books are to balance. We now assume that the recipient is in the nature of a magnitude of volume, and that the recipient input involves increasing this volume. More specifically, we might imagine a river, where the recipient input consists of regulating the flow of water. k can then represent litres of water per time-unit. The more litres of water, the more diluted will the discharge be, and the less the discharges will register on our environmental indicator function. We are still assuming that it is only the sum of discharge that is significant. The environmental indicator function would then look something like this:

123

$$r(\,\cdot\,) = r(-\sum_{j=1}^{n} d_j/k) \tag{6.39}$$

Once again we presuppose that the environmental indicator function is a monotonic increasing function, which gives us the following:

$$\frac{\partial r}{\partial d_j} = -\frac{r'}{k}, \quad \frac{\partial r}{\partial k} = r' \cdot \sum_{j=1}^{n} d_j/k^2, \quad r' > 0 \tag{6.40}$$

Insertion in the passus equation yields:

$$\varepsilon \bar{R} = \sum_{j=1}^{n} \frac{\partial r}{\partial d_j} d_j + \frac{\partial r}{\partial k} k$$

$$= r'(-\frac{1}{k}\sum_{j=1}^{n} d_j + \frac{1}{k}\sum_{j=1}^{n} d_j) = 0 \tag{6.41}$$

We can see here that in this example the passus coefficient will be constant and equal to 0. This implies a balanced budget. The fact that the passus coefficient is constant and equal to 0 means that our environmental indicator function is homogeneous of degree 0 in discharge and recipient input.

Example 3. Tidying up with an exogenous discharge component. The books of the recipient authority may show a deficit. Let us now assume that we have the same situation as in Example 1, with this difference, that we now have a discharge component which the environmental protection authority is unable to influence. In the case of air pollution, for example, we may be dealing with sulphuric oxides conveyed through the air from abroad and incapable of being influenced, whereas national discharges are under the control of environmental protection authorities. We can specify this by stating that the first component of the discharges, d_1 is a positive magnitude which cannot be influenced by the recipient authorities, whereas the other discharges from subscript 2 to n can be influenced. The environmental indicator function will then be:

$$r(\cdot) = r(-d_1 - \sum_{j=2}^{n} d_j + ak) \tag{6.42}$$

The constraint that now applies will be:

$$ak \leqq d_1 + \sum_{j=2}^{n} d_j \tag{6.43}$$

It will be noted that we can now remove more pollutants than those corresponding to what is discharged by the firms over which the environmental protection authorities have control. The optimal budget, when use is made of (6.20) and (6.31), will then be:

$$b = \sum_{j=2}^{n} t^0 d_j^0 - p_k k \tag{6.44}$$

$$= \sum_{j=2}^{n} -\lambda_1 \frac{\partial r}{\partial d_j} d_j - \lambda_1 \frac{\partial r}{\partial k} k$$

$$= -\lambda_1 \left(\sum_{j=2}^{n} \frac{\partial r}{\partial d_j} d_j + \frac{\partial r}{\partial k} k \right)$$

$$= -\lambda_1 \{ r'(-\sum_{j=2}^{n} d_j += ak) \}$$

In order to arrive at the last expression above we have made use of equation (6.36). We get the following result for the size of the budget:

$$b > 0 : ak < \sum_{j=2}^{n} d_j$$

$$b = 0 : ak = \sum_{j=2}^{n} d_j$$

$$b < 0 : ak > \sum_{j=2}^{n} d_j$$

The budget will show a deficit if the position is such that it would be optimal to tidy up a larger amount of discharges than that corresponding to the total discharge from the units over which the environmental protection authorities exercise control.

In practice the desire to balance the budget in such recipient bodies often exists. If a surplus is permitted, this offers no problems in the situations in which reality is of the kind shown in Example 1, that is, that the books are expected to show a surplus. But in situations in which the budget will show a deficit, no superior authority can at the same time insist on a balance, when the optimal charges are specified. It will be impossible for the recipient body to satisfy both conditions at the same time if optimal amounts of discharge are to be realised. This is easily understood if we consider that the deficit can be reduced only by increasing the amount of discharge received or by reducing the recipient input. But in optimum, however, the environmental indicator constraint should apply. By increasing discharge and/or reducing the recipient input, this restriction will no longer be fulfilled. If the demand for a non-negative budget is a real constraint, this constraint must be taken into consideration in calculating optimal charges. This means that we must return to the problem as set out in section (C) Recipient measures. Taking into consideration an extra side condition of this kind is referred to in literature as 'the second-best problem'. We must now consider how the firms adjust to the charge. Let us consider equation (6.29), which tells us that, for an internal solution, the marginal purification costs must equal the charge. We can insert this behavioural equation in the demand for a non-negative budget:

$$\sum_{j=1}^{n} t_j d_j - p_k k = \sum_{j=1}^{n} c'_{pj} d_j - p_k k \geq 0$$

Introducing this side condition in the Lagrange function for the problem in section (C) with the Lagrange parameter λ_4 gives us the following necessary first order conditions:

$$\frac{\partial L}{\partial x_j} = p_j - c'_{xj} + \lambda_2 z'_j \leqq 0 \tag{6.45}$$

$$\frac{\partial L}{\partial z_j} = - c'_{pj} + \lambda_{2j} + \lambda_{3j} + \lambda_4 \, c''_{pj} d_j \leqq 0 \tag{6.46}$$

$$\frac{\partial L}{\partial d_j} = c'_{pj} + \lambda_1 \frac{\partial r}{\partial d_j} - \lambda_{3j} + \lambda_4 \, (- c''_{pj} d_j + c'_{pj}) \leqq 0 \tag{6.47}$$

$$\frac{\partial L}{\partial k} = - p_k + \lambda_1 \frac{\partial r}{\partial k} - \lambda_4 p_k \leqq 0 \tag{6.48}$$

We shall concentrate on the interior solution, that is, where the equations hold with equality, and we also presuppose that the situation does not arise where factories carry out no purification whatsoever. This implies, as previously mentioned, that λ_{3j} will be equal to 0. In interpreting the conditions we must remember that the equation (6.29) must apply throughout. If we consider the condition for adjustment of primary residual generation, equation (6.46), we shall see that if the budget condition is binding, that is, λ_4 is positive, this means that the charge will now be greater than the shadow price for primary generation of residuals. The second derivative of the purification cost function is by definition positive. As far as the condition for adjustment of residual discharge is concerned, we now have a situation in which the charge may be both bigger and smaller than the marginal damage. From (6.29) and when $\lambda_{3j} = 0$, (6.47) can be written

$$t_j - (- \lambda_1 \frac{\partial r}{\partial d_j}) = - \lambda_4 (- c''_{pj} d_j + c'_{pj}) \tag{6.49}$$

The sign of the left-hand side will depend on the term inside the bracket for λ_4 in equation (6.32). We can rewrite this in order to arrive at a clearer interpretation:

$$-c''_{pj} d_j + c'_{pj} = c'_{pj} (1 + \check{c}'_{pj}) \quad \text{where } \check{c}'_{pj} = \frac{-c''_{pj} d_j}{c'_{pj}} < 0$$

The elasticity of marginal costs in purification with respect to

127

residual discharge is negative. Equation (6.49) tells us that the charge will be greater than the marginal damage of residual discharge, if the numerical value of the elasticity is greater than 1, and correspondingly it will be less than the marginal damage if the numerical value of the elasticity of the purification marginal cost function, with respect to discharge, is less than 1. As far as adjustment of recipient input is concerned, we can see from equation (6.48) that with a binding budget constraint the recipient activity will be applied in such a way that the price per unit is lower than the marginal benefit we obtain from the recipient input.

Comparing this solution with the previous optimal solution is no easy matter. The main point, however, is clear: in introducing a demand for a balanced budget in situations where this is binding, this demand will influence and distort the whole solution for discharge and input of recipient measures. In other words, a social cost is involved in introducing demands for a budget balance of this nature. But in a situation involving a deficit for a recipient body this, too, has to be financed. If we introduce a larger model, we can say that there might also be negative effects in financing deficits of this nature via the tax system or in other ways. It is not possible to state in advance where the demand for financial coverage will inflict least damage.

6.3 A GENERAL EQUILIBRIUM MODEL

In the various model versions in section 6.2 we put a price indirectly on the environment in the form of the shadow price that emerged on the environmental indicator constraint. We shall now pose a more fundamental question as to whether it would be possible to arrive at a price for the environment. For this purpose we shall, of course, require a more general model than in the previous section, and we shall consider a general equilibrium model. We shall not be introducing externally given prices, as was the case in section 6.2. Among the advantages of a general equilibrium model of this kind may be mentioned the fact that it is suitable for providing an overall view of the complexity of the problem. It enables us to chart mutual relations and to investigate the weaknesses of partial solutions. A model analysis, too, may prove useful in establishing concepts or arriving at useful variables, on which further empirical work can be carried out. The analysis will show what sort of information has to be gathered, in order to be able to deal with the allocation problem. A general equilibrium model can also provide a starting point for an investigation of the possibility of decentralising decisions and examining the efficiency of the various means. Let us consider a model consisting of the following six equations:

$$F(x_1, \ldots, x_n, k) \leqq 0, \quad \frac{\partial F}{\partial x_j} > 0, \quad \frac{\partial F}{\partial k} > 0 \tag{6.50}$$

$$z_j(x_j) = z_j, \quad j=1, \ldots, n, \quad \frac{\partial z_j}{\partial x_j} > 0 \tag{6.51}$$

$$z^i(x_1^i, \ldots, x_n^i) = z^i, \quad i=1, \ldots, m, \quad \frac{\partial z^i}{\partial x_j^i} > 0 \tag{6.52}$$

$$U^i = U^i(x_1^i, \ldots, x_n^i, R), \quad \frac{\partial U^i}{\partial x_j^i} > 0, \quad \frac{\partial U^i}{\partial R} > 0 \tag{6.53}$$

$$r(z_1, \ldots, z_n, z^1, \ldots, z^m, k) = R, \tag{6.54}$$

$$\frac{\partial r}{\partial z_j} < 0, \quad \frac{\partial r}{\partial z^i} < 0, \quad \frac{\partial r}{\partial k} > 0$$

$$\sum_{i=1}^{m} x_j^i \leqq x_j \tag{6.55}$$

Equation (6.50) is a transformation function. It shows in what least cost combination the n goods $x_1, \ldots, x_n$ and purification activities, k, can be produced with given resources. If equation (6.50) applies with inequality, this means that the given resources are not

129

fully utilised. Equation (6.51) shows the generation of primary residuals from the production of every single good. This equation corresponds to equation (6.52) in section 6.2. When the goods x_1, ..., x_n are consumed by the consumers, residuals may also arise. This is expressed in equation (6.52), where it is presupposed that all the goods which consumer number i consumes result in a joint discharge of residuals. As an example we might imagine the intake of various types of food and drink producing a common organic discharge. A more complicated version would be residuals to be linked with the consumption of every single good separately. Equation (6.53) shows consumers' utility functions, that is, how they evaluate the various goods and the environmental quality R. We assume that it is beyond the control of the individual consumer to determine the quality of the environment. Equation (6.54) is the environmental indicator function. Here we have a more complicated formula than in the earlier sections, because we are here dealing with discharges both from production of goods and from the consumption of goods. Equation (6.55) is an equilibrium condition, that is, the total consumption of a type of good j has to be less than or equal to the production of this good. The utility functions (6.53) are qualitatively new equations inserted in the model as compared with section 6.2.

What should the objective function in our problem? A traditional formulation from the turn of the century is to maximise the utility for *one* person, while the utility for the other persons remains constant. This reflects the so-called Pareto principle. This principle states that we can achieve an improvement in society if the utility for one person can be increased, without the utility for any other person being reduced. Pareto optimum will occur when it is not possible to increase the utility for one person without reducing that of at least one other person. It is, of course, coincidental what person we select as the one whose utility is to be maximised. We can therefore just as conveniently introduce a symmetrical objective function in the following way:

$$\sum_{i=1}^{n} w^i U^i(x_1^i, \ldots, x_n^i, R) \tag{6.56}$$

where w^i represents positive magnitudes which convert the separate 'util' measurements with which we might measure utility to a

130

common unit of measure. Let us call the common unit of measure an 'administrative unit'. A single magnitude or weight w^i could then be called an administrator unit per Mr. Inglis util. Let us now suppose that we are considering changes in the utility for Mr Inglis (number i), and Mr Jones (number j). The total utility sum, measured in administrative units, is to remain constant. Differentiating equation (6.56) then gives us

$$\mathrm{d}U^i / \mathrm{d}U^j = w^j / w^i \tag{6.57}$$

This relationship expresses the reduction we have achieved in the Inglis utility, when Jones's utility has increased by one unit. If Inglis was now the person we have selected with a view to maximising his utility, while the constant utility levels of all the other persons were introduced as side conditions, we shall see that the weight ratio in (6.57) corresponds to the shadow prices we would have for these side conditions for constant utility levels. Use of (6.56) as an objective function is therefore equivalent to maximising the utility of one person, while all the other utilities are kept constant. Relation (6.56), however, can also provide further interpretation. We can consider (6.56) as a so-called welfare function for our social administrator. It is quite simple, consisting merely of a linear summation of the utility levels of individual persons. We are presupposing that all weights are positive. The welfare function (6.56) belongs to the class of what we call Bergson welfare functions of the Pareto type.

The pollution problem in our model consists in discharge from production of goods and from the consumption of goods influencing the quality of the environment, which in turn enters into consumers' utility functions. The more discharge, the lower the value recorded by our environmental indicator and the lower the utility enjoyed by each individual person. For the sake of simplicity we consider all discharges as one-dimensional magnitudes, that is, discharges from both production and the consumption of a good have the same physical denomination and are one-dimensional. Our environmental indicator function is also treated as one-dimensional. An extension in this case is quite straightforward. Note that in this section purification within the firms and households is ignored so that primary waste is equal to the amounts discharged.

131

The optimal discharge and the environmental quality may be found by solving the following problem:

$$\text{Maximise}_{x_j, x_j^i, z_j, z^i, R, k} \sum_{i=1}^{m} w^i U^i(x_1^i, \ldots, x_n^i, R)$$

subject to

$$F(x_1, \ldots, x_n, k) \leq 0$$
$$z_j(x_j) = z_j$$
$$z^i(x_1^i, \ldots, x_n^i) = z^i$$
$$r(z_1, \ldots, z_n, z^1, \ldots, z^m, k) = R$$
$$\sum_{i=1}^{m} x_n^i \leq x_j$$

The Lagrange function for our problem will be as follows:

$$L = \sum_{i=1}^{m} w^i U^i(x_1^i, \ldots, x_n^i, R) \tag{6.58}$$

$$- \lambda_1 F(x_1, \ldots, x_n, k)$$

$$- \sum_{j=1}^{n} \lambda_{2j} (z_j(x_j) - z_j)$$

$$- \sum_{i=1}^{m} \lambda_{3i}(z^i(x_1^i, \ldots, x_n^i) - z^i)$$

$$- \lambda_4(R - r(z_1, \ldots, z_n, z^1, \ldots, z^n, k))$$

$$- \sum_{j=1}^{n} \lambda_{5j}(\sum_{i=1}^{m} x_j^i - x_j)$$

The necessary first order conditions are:

$$\frac{\partial L}{\partial x_j} = -\lambda_1 \frac{\partial F}{\partial x_j} - \lambda_{2j} \frac{\partial z_j}{\partial x_j} + \lambda_{5j} \leq 0 \tag{6.59}$$

$$\frac{\partial L}{\partial x_j^i} = w^i \frac{\partial U^i}{\partial x_j^i} - \lambda_{3i} \frac{\partial z^i}{\partial x_j^i} - \lambda_{5j} \leq 0 \qquad (6.60)$$

$$\frac{\partial L}{\partial z_j} = \lambda_{2j} + \lambda_4 \frac{\partial r}{\partial z_j} \leq 0 \qquad (6.61)$$

$$\frac{\partial L}{\partial z^i} = \lambda_{3i} + \lambda_4 \frac{\partial r}{\partial z^i} \leq 0 \qquad (6.62)$$

$$\frac{\partial L}{\partial R} = \sum_{i=1}^{m} w^i \frac{\partial U^i}{\partial R} - \lambda_4 \leq 0 \qquad (6.63)$$

$$\frac{\partial L}{\partial k} = -\lambda_1 \frac{\partial F}{\partial k} + \lambda_4 \frac{\partial r}{\partial k} \leq 0 \qquad (6.64)$$

In addition, the Lagrange multipliers, multiplied by the relevant constraints, are to be equal to 0. We shall concentrate on an interior solution, that is, with equations (6.59)–(6.64) holding with equality. If an equation cannot be fulfilled with equality, this means that the relevant variable is set equal to 0.

Equation (6.59) shows the adjustment of production in the case of good j. The first term expresses the cost of production. By cost in this case is meant alternative cost: a unit increase of good j will necessitate a reduction in the amount of other goods, as we have constant resources. The second term measures the cost of a deteriorated environmental quality caused by an increase of primary residuals. The third term expresses the utility of good j. λ_{5j} is the shadow price of increasing the total supply to consumers of good number j. The sum of the marginal cost components must be equal to the marginal utility of good number j. Equation (6.60) shows us how the marginal utility of good number j is determined. The first component expresses the private marginal utility of good j for person i, weighted by w^i. We shall call this the socially weighted private marginal utility.

133

The other term expresses the cost that arises when consumption involves a discharge of residuals. The marginal utility of good j is equal to the socially weighted private marginal utility minus the environmental cost component. We note that this difference is to be equal for all persons. The shadow price λ_{5j} has no subscript for persons. Equations (6.61) and (6.62) determine the shadow prices for the generation of discharge from the production of good number j and consumption respectively. The shadow price for the environmental indicator constraint λ_4 is determined in equation (6.63). We see that the shadow price is equal to the total sum of all the marginal utility increases arising from an increase of one unit in the environmental quality. These private utility increases are weighted by the weights w^i. The interpretation of the shadow price λ_4 is that it shows the change in objective function resulting from a change of one unit in the constraint. In this case the constraint is the environmental indicator R. We note that we get an increase in the objective function, which is the weighted sum of all utility functions, when the environmental quality rises by one unit. Since the environmental quality appears as a *public good,* that is, it is an integral part of all persons' utility functions, the result is that the marginal effect of an increase in R must be a sum of all marginal effects for all persons.

A public good can be defined in various ways: a particular feature of the utility function specified in (6.53) is that all the persons in our economy are faced with the same environmental quality. No one can be supplied with a higher environmental quality without all other persons, too, having access to this improved environmental quality. No person can be excluded or debarred from enjoying the environmental quality. Private goods and services and the environmental quality are separated by a semicolon, indicating that individual persons cannot affect the magnitude of the environmental quality. Environmental quality is a public good; the x's are private goods. The individual determines himself what amounts he wishes to consume of these latter goods.

To get a better overall idea of what the solution (6.59)–(6.64) involves, we can eliminate the Lagrange multipliers. If we add together (6.59) and (6.60), λ_{5j} is eliminated. We substitute for λ_{2j} and λ_{3i} from equations (6.61) and (6.62) and finally remove λ_4 by insertion from equation (6.58). The adjustment of consumption of good j for person number i and the production of good j can then be described in the following manner:

$$w^i \frac{\partial U^i}{\partial x_j^i} + \sum_{s=1}^{m} w^s \frac{\partial U^s}{\partial R} \left(\frac{\partial r}{\partial z^i} \frac{\partial z^i}{\partial x_j^i} + \frac{\partial r}{\partial z_j} \frac{\partial z_j}{\partial x_j} \right) = \lambda_1 \frac{\partial F}{\partial x_j} \qquad (6.65)$$

| private socially weighted marginal utility | total marginal environmental costs via residuals from consumption and production | cost in production |

The second term on the left-hand side in (6.55) shows the value of the external effects in our model. Inside the parentheses the first term is the effect we get on the environmental indicator function when increased consumption of good j for person number i generates a discharge from person number i. This discharge will in turn affect the level of the environmental indicator index. The second term represents the discharge that production of good j produces. This discharge will then in turn affect the environmental indicator index. These effects are both measured in the same physical unit. Remember that $\partial r/\partial z_j$ and $\partial r/\partial z^i$ are both negative. How is the change evaluated in the units in which the objective function is measured? Evaluation is based on the marginal effect on the utility level of every individual due to the same change in the environmental indicator. These different utility measurements are then brought on the same measuring unit again, using the weight system. The denomination for the entire left-hand side is therefore the administrative unit per physical unit of x_j^i, for example, kilos. This is the same denomination as on the right-hand side, as the shadow price λ_1 converts the physical denomination of the derivative of the transformation function to the denomination of the objective function. We can say that the entire left-hand side expresses the social marginal utility resulting from a good being produced and consumed by person number i. The evaluation of the pollution effects constitutes the difference between the private and social marginal utility of the good.

The shadow price λ_1 can be eliminated if we consider the adjustment of two goods for person number i. If we consider the corresponding condition as in (6.65) for good number d, by dividing the expressions on each side of the equality sign by one another, we get the following:

$$\frac{\dfrac{\partial U^i}{\partial x_j} + \sum\limits_{s=1}^{m} \dfrac{w^s}{w^i} \dfrac{\partial U^s}{\partial R} \left(\dfrac{\partial r}{\partial z^i} \dfrac{\partial z^i}{\partial x_j} + \dfrac{\partial r}{\partial z_j} \dfrac{\partial z_j}{\partial x_j} \right)}{\dfrac{\partial U^i}{\partial x_d} + \sum\limits_{s=1}^{m} \dfrac{w^s}{w^i} \dfrac{\partial U^s}{\partial R} \left(\dfrac{\partial r}{\partial z^i} \dfrac{\partial z^i}{\partial x_d} + \dfrac{\partial r}{\partial z_d} \dfrac{\partial z_d}{\partial x_d} \right)} = \frac{\dfrac{\partial F}{\partial x_j}}{\dfrac{\partial F}{\partial x_d}} \tag{6.66}$$

Above we have arranged the left-hand side in such a way that only relative weight ratios are introduced, namely, that all persons' weights are seen in relation to the weight for person number i. This condition will then be entirely analogous with the one we would have had if, in tackling our problem, we had chosen to maximise the utility level for person number i, with constant utility levels for all other persons as a side condition. The right-hand side in (6.66) is called the marginal rate of transformation between good j and good d. It shows by how many units the production of good d will have to be reduced, when the production of good j increases by one unit. The relationship between marginal utilities for two different goods for the same person is called in general the marginal rate of substitution. It measures what reduction a person must make in the amount of one good when the amount of the other good increases by one unit, with the utility level remaining constant. The entire left-hand side of (6.65) is called the marginal social rate of substitution for goods j and d for consumer number i. We see that the marginal social rate of substitution must be equal for all individuals for the same pair of goods, whereas the private marginal rate of substitution may differ from one individual to another. This will depend on the technical pollution relationships. In the first place it is a question of whether the discharge technology is the same or different from one person to another for the same good. In the second place it is a question whether discharge measured in the same unit from various persons has a different effect on the environmental indicator function. If people have the same technology in the generation of residuals, and the effect on the environmental indicator function is independent of the person responsible for the discharge, the private marginal rate of substitution for the same pair of goods will also be equal from one person to another.

We shall now consider how the optimal solution can be implemented. As far as consumers are concerned every single consumer is assumed to maximise his utility on the basis of a budget constraint.

Every single consumer is faced with given prices for the various goods, p_j, and he considers the charge on generation of residuals, t^i, and his income, y^i, as given magnitudes.

Consumer number i's problem is

Maximise $U^i(x^i_1, \ldots, x^i_n, R)$
$x^i_1, \ldots, x^i_n, z^i$

subject to

$$\sum_{d=1}^{n} p_d x^i_d + t^i z^i \leq y^i$$

$$z^i(x^i_1, \ldots, x^i_n) = z^i$$

$$x^i_d, z^i \geq 0$$

The Lagrange function for a typical consumer's maximisation problem will then be:

$$L = U^i(x^i_1, \ldots, x^i_n, R) \tag{6.67}$$

$$- \mu^i_1 \left(\sum_{d=1}^{n} p_d x^i_d + t^i z^i - y^i \right)$$

$$- \mu^i_2 \left(z^i(x^i_1, \ldots, x^i_n) - z^i \right)$$

The consumer must also take into account his discharge technology (6.52) in making his adjustment. The consumer can reduce his discharge by reducing his consumption of one or several goods. The necessary first order conditions will be

$$\frac{\partial L}{\partial x^i_j} = \frac{\partial U^i}{\partial x^i_j} - \mu^i_1 p_j - \mu^i_2 \frac{\partial z^i}{\partial x^i_j} \leq 0 \tag{6.68}$$

$$\frac{\partial L}{\partial z^i} = - \mu^i_1 t^i + \mu^i_2 \leq 0 \tag{6.69}$$

We shall concentrate on an interior solution. If we eliminate the Lagrange multiplier μ_2, we get

137

$$\frac{1}{\mu_1^i} \frac{\partial U^i}{\partial x_j^i} = p_j + t^i \frac{\partial z^i}{\partial x_j^i} \qquad (6.70)$$

On the left-hand side we have the marginal utility of good j for person number i, weighted with the inverse value of the Lagrange multiplier μ_1^i. It expresses the shadow price of the budget constraint, that is, the increase in the objective function, which is here the utility function, when income increases by one unit. This magnitude is often called the marginal utility of money. It is described in terms of utils per pound sterling. The marginal utility is designated in utils per physical unit of x_j, for example, kilos, so that the entire left-hand side is designated in pounds sterling per kilo. This is also the denomination on the right-hand side.

As far as producers are concerned we have not yet included the way in which they adjust the use of raw materials and resources such as labour, intermediary goods, etc. The individual firms are not identified in the transformation function. For this reason we assume that the production side is treated as an aggregated decision unit. As an objective function we then have a profit function consisting of total sales earnings minus charges on residuals. We assume that the aggregated production sector interprets the product prices p_j and the charge t_j as given magnitudes. The problem of the production sector is

$$\underset{x_1, \ldots, x_n, z_1, \ldots, z_n}{\text{maximise}} \sum_{j=1}^{n} (p_j x_j - t_j z_j)$$

subject to

$$F(x_1, \ldots, x_n) \leqq 0$$
$$z_j(x_j) = z_j \quad j = 1, \ldots, n$$
$$x_j, z_j \geqq 0$$

The Lagrange function for the maximisation problem of the aggregated production sector will be

$$L = \Sigma_j \, (p_j x_j - t_j z_j) \qquad (6.71)$$
$$\quad - \gamma_1 F(x_1, \ldots, x_n)$$

$$- \sum_{j=1}^{n} \gamma_{2j}(z_j(x_j) - z_j)$$

The necessary first order conditions are

$$\frac{\partial L}{\partial x_j} = p_j - \gamma_1 \frac{\partial F}{\partial x_j} - \gamma_{2j} \frac{\partial z_j}{\partial x_j} \leq 0 \qquad (6.72)$$

$$\frac{\partial L}{\partial z_j} = - t_j + \gamma_{2j} \leq 0 \qquad (6.73)$$

Elimination of the shadow price γ_{2j}, when we have an interior solution, yields:

$$p_j = \gamma_1 \frac{\partial F}{\partial x_j} + t_j \frac{\partial z_j}{\partial x_j} \qquad (6.74)$$

The production sector adjusts production of good number j so that the product price of this good corresponds to the alternative costs in the production of the good, plus the cost due to generation of residuals in the production of good j. In our market system we presuppose that consumers and producers are faced with the same prices. Provided that there exists an unambiguous internal solution for our consumer and producer problem, we can combine the adjustment conditions on the consumer and production sides:

$$\frac{1}{\mu_1^i} \frac{\partial U^i}{\partial x_j^i} - t^i \frac{\partial z^i}{\partial x_j^i} - t_j \frac{\partial z_j}{\partial x_j} = \gamma_1 \frac{\partial F}{\partial x_i} \qquad (6.75)$$

This relation expresses the fact that the marginal utility for good j for person number i, measured in pounds sterling per kilo, minus the amount of the charge for generation of residuals, which derives from a marginal increase in the consumption of good number j, minus the amount of the charge for the generation of residuals in the production of good j, is to be equal to the opportunity costs for the production of good j. If we now compare equation (6.75) with equation (6.65), we shall see that the inverse value of the marginal utility of money here plays the

139

same role as the social weight in (6.65).

Furthermore, the counterpart to t^i is:

$$\sum_{s=1}^{m} w^s \frac{\partial U^s}{\partial R} \frac{\partial r}{\partial z^i}$$

and for t_j:

$$\sum_{s=1}^{m} w^s \frac{\partial U^s}{\partial R} \frac{\partial r}{\partial x_j}$$

We note that all magnitudes in equation (6.75) are measured in pounds sterling per kilo, whereas all magnitudes in equation (6.65) are measured in administrative units per kilo. In other words, we cannot state that the marginal utility of money is to be the equivalent of a social weight, and that the two charges are to be equal to the expressions we have just mentioned. Let us consider an equation corresponding to equation (6.66).

$$\frac{\dfrac{\partial U^i}{\partial x_j} - \mu_1^i t^i \dfrac{\partial z^i}{\partial x_j} - \mu_1^i t_j \dfrac{\partial z_j}{\partial x_j} \quad \dfrac{\partial F}{\partial x_j}}{\dfrac{\partial U^i}{\partial x_d^i} - \mu_1^i t^i \dfrac{\partial z^i}{\partial x_d^i} - \mu_1^i t_d \dfrac{\partial z_d}{\partial x_d} \quad \dfrac{\partial F}{\partial x_d}} = \overline{} \tag{6.76}$$

The Lagrange multiplier γ_1 is eliminated by combining the adjustment condition for production and consumption of goods numbered d and j for person number i. Let us go back to (6.66) in order to see the connection with (6.76) and consider the denomination of the magnitude shown in the numerator on the left-hand side in (6.66). The first term is measured in utils per kilo. The next term looks a little more complicated, but if we start inside the bracket and work our way out, what we get is that the denomination of the partial derivatives of the generation function must be residual units per kilo of x_j. The denomination of the partial derivatives of the environmental indicator function must be environmental index units per residual units. The denomination of the partial derivatives of the utility function with regard to our environmental indicator must be s-utils per environmental index units. The relative weight relationships w^s/w^i is designated i-util per s-util. All these relative weight relations convert all other persons' util measurements to i-person util measurements. The net result is that we are left with the entire total expression having as its denomination i-utils per kilo x_j. If we now consider the magnitude of the numerator on the left-hand side in equation (6.76), we shall find that the first term naturally is measured in i-utils per kilo x_j. The next two terms are also measured in i-utils per kilo x_j. This may be seen as the marginal utility of money has the denomination i-util per pound sterling, the charge is denominated in pounds sterling per residual from person i, and the partial derivatives of the generation function for person number i are designated residual units per kilo x_j. If these are multiplied together, we are left with i utils per kilo x_j. We can see that the right-hand side in (6.76) corresponds to the right-hand side in (6.66). Thus the elements in (6.76) are directly comparable with those in (6.66).

It is important to note here that the absolute price and income level plays no role. The absolute charge level is also arbitrary. When the prices are given, the absolute level for charges follows. They must be such that the optimal solution described in (6.66) is realised. A comparison of (6.66) and (6.76) shows that it is possible to effect the optimal solution in a market system with the introduction of a charge. It should be emphasised that this is not necessarily so, but merely that it is possible. The question whether it must be so, and what conditions must be fulfilled in order that it must be so, is very complex, and will not be dealt with here. In order that the optimal solution is to be realised, the relationship between the marginal utility of money for person number i and person number s must be equal to the weight relationship w^s/w^i. A weight relationship of this nature, as already mentioned, is designated in i-utils per unit s-util. The relationship between the marginal utility of money is also given precisely this denomination, as μ_1^i is i-utils per pound sterling and μ_1^s is s-utils per pound sterling. This implies that the product of the marginal utility of money, e.g. μ_1^i, and the weight corresponding to this person w^i, must be constant and independent of i. If we now consider the solution for charges, we get the following:

$$t^i = -\frac{1}{\mu_1^i} \sum_{s=1}^{m} \frac{w^s}{w^i} \frac{\partial U^s}{\partial R} \frac{\partial r}{\partial z^i} \tag{6.77}$$

$$t_j = -\frac{1}{\mu_1^i} \sum_{s=1}^{m} \frac{w^s}{w_i} \frac{\partial U^s}{\partial R} \frac{\partial r}{\partial s_j} \tag{6.78}$$

The charge on effluents from consumers will vary from one person to another, assuming they have different residual technologies. If they possess the same technology, they will face the same charge. With regard to the charge on effluents from the production of good j, we can see from (6.78) that this is naturally independent of the personal numbering since the product $\mu_1^i \, w^i$ is constant and independent of i.

With regard to the use of effluent charges, we can see that their basis is the discharge of residuals. We are therefore presupposing that discharge can be measured unambiguously. In order to calculate the rate of charge we must know in the first place how the discharge affects the environmental indicator, and in the second place we must calculate how a change in the environmental quality is evaluated as a collective good. In our model we have taken the individual utility functions as a starting point for our calculation, so that the appraisal of collective goods arises as the sum of the marginal evaluations of every single individual. If we take as our starting point a welfare function with all the individual utility functions as arguments, we can state that the marginal utilities of environmental quality are weighed with the weights of the welfare function. The absolute level for charges is *per se* arbitrary in an equilibrium model of this kind. The absolute charge level, therefore, will depend on the absolute price and income levels.

When it comes to determining the level of k, we can, in the same way as in section 6.2, establish a recipient body whose task involves collecting charges from any person making a discharge, that is in this case both producers and consumers, and also responsible for implementing measures in the recipient. We also assume the same institutional set-up as in the foregoing section. If we have a superior environmental protection authority, with full information, solving our problem in the form of solutions from (6.59) to (6.64), and given that consumers and producers behave as presupposed above, the optimal charges can be calculated and delegated to the subordinate recipient authority. As we now have no restriction on the level of the environmental indicator, an evaluation of the environmental quality must be introduced into the objective function. We can, for example, imagine that the subordinate recipient authority is given a price per unit of environmental indicator. The local authority is now entrusted with the task of maximising a surplus which is given by the income from the sale of environmental services plus all charge revenue, minus the expenses of the recipient activity. The sale of environmental services is a mere desk operation or an accountancy problem, as the environmental quality is a collective good, and per definition this cannot be apportioned out and sold in the concrete sense of the word. The problem can be postulated as follows:

$$\text{Maximise}_{z^i, z_j, R, k} \ p_R R + \sum_{i=1}^{m} t^i z^i + \sum_{j=1}^{n} t_j z_j - p_k k$$

subject to

$$r(z^1, \ldots, z^m, z_1, \ldots, z_n, k) = R$$
$$R, z^i, z_j, k \geq 0$$

The Lagrange function for this problem is

$$L = p_r R + \sum_{i=1}^{m} t^i z^i + \sum_{j=1}^{n} t_j z_j - p_k k \qquad (6.79)$$

$$- \lambda (R - r(z^1, \ldots, z^m, z_1, \ldots, z_n, k))$$

The necessary first order conditions are

$$\frac{\partial L}{\partial z^i} = t^i + \lambda \frac{\partial r}{\partial z^i} \leq 0 \qquad (6.80)$$

$$\frac{\partial L}{\partial z_j} = t_j + \lambda \frac{\partial r}{\partial r_j} \leq 0 \qquad (6.81)$$

143

$$\frac{\partial L}{\partial R} = p_R - \lambda \leq 0 \tag{6.82}$$

$$\frac{\partial L}{\partial k} = -p_k + \lambda \frac{\partial r}{\partial k} \leq 0 \tag{6.83}$$

We shall concentrate on an interior solution. Let us now see whether the optimal solution can be realised with this kind of set-up. The price p_R, which the superior authority must give to the local recipient body as a price for the environment, may be found from equation (6.65). The effluent charges for respectively consumers, t_i, and producers, t_j, may be found in the first round from equations (6.62) and (6.61). We must constantly bear in mind that the charge level is arbitrary, which means that we ignore a proportionality factor when we determine the charges in this way. If we look at the market solution in equations (6.70) and (6.74) and the optimal solutions (6.59)–(6.64), and our solutions here, (6.80)–(6.83), we shall see that this institutional arrangement may produce the optimal result. The price of goods here will then have to correspond to the opportunity costs we operated with in equations (6.59)–(6.64).

With regard to the question of the financial result for the local recipient body, this will be entirely analogous with our analysis previously. The sum of the charge revenues minus expenses for recipient input may be called the financial surplus. We do not, in fact, introduce the total value of environmental services here. Inserting for all charges and prices from equations (6.80)–(6.83), yields:

$$b = \sum_{i=1}^{m} t^i z^i + \sum_{j=1}^{n} t_j z_j - p_k k \tag{6.84}$$

$$= -\sum_{i=1}^{m} \lambda \frac{\partial r}{\partial z^i} \cdot z^i + \sum_{j=1}^{n} -\lambda \frac{\partial r}{\partial z_j} z^j - \lambda \frac{\partial r}{qk} k$$

$$= -\lambda \, \varepsilon \, R$$

ε once again stands for the passus coefficient in the environmental indicator function. The financial budget is positive,

144

equal to zero or negative according to whether the passus coefficient is negative, equal to zero, or positive. We also note that this result is not affected by the price imputed to the environment, provided that the passus coefficient does not change its sign with a variation in price, as the passus coefficient and the level of the environmental quality will vary with the variation in price p_R, which, of course, is equal to λ.

REFERENCES AND FURTHER READING

Bohm, P. (1967): 'External economics in production', *Stockholm Economic Studies*, Almqvist & Wiksells, Uppsala

Bolwig, N.G. (1971): 'A survey of the economic theory of pollution', *Nationaløkonomisk Tidsskrift*, hefte 3-4, 179-207

Boulding, K.E. (1972): 'Discussion' (Knese, 1971b), *The American Economic Association*, papers and proceedings, 61 (2), 167-9

Coase, R. (1960): 'The problem of social cost', *Journal of Law and Economics*, 3, 1-44

Fisher, A.C. (1981): *Resource and environmental economics*, Cambridge University Press, Cambridge

Frisch, R. (1965): *Theory of Production*, D. Reidel, Dordrecht, Holland

Førsund, F.R. (1971): 'Allocation in space and environmental pollution', *The Swedish Journal of Economics*, 74 (1), March, 19-34

Førsund, F.R. (1974): 'Discussion' (of the paper by Serge-Christophe Kolm), *The management of water quality and the environment*, J. Rothenberg and Ian G. Heggie (eds.), Macmillan, London, pp. 177-88

Førsund, F.R. (1975): 'Spatial aspects of environmental pollution: a public goods approach', *Dynamic allocation of urban space*, A. Karlqvist, L. Lundqvist and F. Snickars (eds.), Saxon House, D.C. Heath, Farnborough, pp. 211-24

Førsund, F.R. and S. Strøm (1973): 3.8 Vedlegg til kap. 3, 'Argumenter for avgifter som virkemiddel i miljøpolitikken', *Langtidsprogrammet*, 1974-7. Spesialanalyse 1. *Forurensninger*, Finansdepartementet, pp. 44-50

Førsund, F.R. and S. Strøm (1974): 'Spillprodukter i den norske økonomien. En makroøkonomisk analyse', *Memorandum* fra Sosialøkonomisk institutt, Universitetet i Oslo, 21. juni

Kolm, S-C. (1974): 'Qualitative returns to scale and the optimum financing of environmental policies', *The management of water quality and the environment*, J. Rothenberg and Ian G. Heggie (eds.), Macmillan, London, 151-71

Kneese, A.V. (1971a): 'Background for the economic analysis of environmental pollution', *The Swedish Journal of Economics*, 73 (1), 1-24

Kneese, A.V. (1971b): 'Environmental pollution: economics and policy', *The American Economic Association*, papers and proceedings, 61 (2), 153-66

Mäler, K.G. (1971): 'A method of estimating social benefits from pollution control', *The Swedish Journal of Economics*, 73 (1), 13-20

Mäler, K.G. (1974): *Environmental economics: a theoretical inquiry*, The Johns Hopkins Press, Baltimore and London

Mäler, K.G. (1985): 'Welfare economics and the environment' in *Handbook of Natural Resource and Energy Economics*, A.V. Kneese and J.L. Sweeny (eds), North Holland, Amsterdam, New York, Oxford, 3-60

Marshall, A. (1890): *Principles of Economics*, Macmillan, London, 1966 (8th edn)

Meade, J.E. (1952): 'External economies and diseconomies in a competitive situation', *Economic Journal*, 62 (2), 54-67

Meade, J.E. (1973): *The theory of economic externalities*, A.W. Sijthoff–Leiden, Institut Universitaire de Hautes Etudes Internationales, Geneva

Mishan, E.J. (1965): 'Reflections of recent developments in the concept of external effects', *Canadian Journal of Economics and Political Science*, 31 (1), February, 4-34

Mishan, E.J. (1971): 'The postwar literature on externalities: an interpretative essay', *Journal of Economic Literature*, 9 (1), 1-28

Mohring, H.D. and M. Harwitz (1962): *Highway benefits: an analytical framework*, Northwestern University Press, Evanston, Illinois

Pearce, D.W. (1976): *Environmental Economics*, Longman, London and New York

Pigou, A.C. (1920): *The economics of welfare*, Macmillan, London (4th edn)

Russell, C.S. and W.O. Spofford, Jr (1972): 'A quantitative framework for residuals management decisions', in *Environmental quality and the social sciences: theoretical and methodological studies*, A.V. Kneese and B.T. Bower (eds), The Johns Hopkins Press, Baltimore and London, 115-79

Sandmo, A. (1972): 'Optimality rules for the provision of collective factors of production', *Journal of Public Economics*, 1, 149-57

Sandmo, A. (1973): 'Public goods and the technology of consumption', *The Review of Economic Studies*, 60 (4), 517-28

7

Pollution and Uncertainty

7.1 INTRODUCTION

In our pollution analyses we have so far considered discharges as deterministic variables. Furthermore, we have assumed that the effects of discharges on environmental indicators can be predicted with certainty. In many actual situations, on the other hand, it would be realistic to assume that both discharges and effects on the environment of given discharges are stochastic variables. Discharges 'measured at the factory gate' may be stochastic on account of random variations in the course of production and/or purifying processes due to mechanical defects, accidents, and the like. Discharges reaching recipients may vary at random as a result of weather conditions. As far as the relationship between discharge reaching the recipient and 'environmental service production' measured in terms of the value of environmental indicators are concerned, these can also be affected by wind and weather conditions. Damage caused by discharge to a river may depend on the water level; wind and precipitation will also have a bearing on the effects in the actual recipient of supplies of given quantities of discharge. An example that springs to mind is the formation of photochemical smog.

In this chapter we shall consider how the introduction of uncertainty, as mentioned above, affects the conclusions in the simplest model in Chapter 6.2. We shall also deal with an entirely different type of uncertainty problem, namely, how does a discharge source (factory) adjust itself when an effluent charge is introduced and reports submitted by the firm itself are controlled by means of random checks?

7.2 STOCHASTIC DISCHARGE AND ENVIRONMENTAL REACTIONS

Let us consider the partial pollution model (6.1)–(6.4) in section 6.2. Production is given in each of the factories. Instead of minimising total purification costs, given a constraint on the environmental indicator, we shall now measure environmental services in terms of money in introducing a function, the damage function, which converts environmental indicator values into monetary units.

Instead of equation (6.4) in section 6.2, the following will now apply:

$$D(\, r(\, \sum_{i=1}^{n} d_i + \sum_{i=1}^{n} v_i, u)),\ D' < 0,\ D'' > 0,\ r_1' < 0,\ (7.1)$$
$$r_2' < 0,\ r_{11}'' < 0,\ r_{12}'' = 0$$

Here d_i is *planned* discharge from factory number 1; $d_i + v_i$ represent realised (actual) discharge from factory number i reaching the recipient; v_i is a stochastic variable; u is a stochastic variable taking care of random variations in the environment's service production capacity; and $D(\cdot)$ is a function converting physical units for the environmental indicator into terms of money. $D(\cdot)$ may be called a damage function; in the event of an increase in discharge it measures the value of lost environmental service production. It is assumed that the damage function is convex. In relation to equation (6.4) in section 6.2, the environmental indicator function has been simplified, so that localisation of the source of discharge plays no role. We assume that marginal reduction of environmental services will be greater in absolute value, the greater the discharge is. As far as the effects of the stochastic variable u are concerned, we have conventionally made the marginal effect negative, and for the sake of simplicity ignored cross-effects ($r_{12}'' = 0$). Typical graphs for the functions are shown in Figure 7.1.

With regard to the individual firm's possibilities for purification, we shall stick to the purification cost functions (6.3) in section 6.2. Discharge d_i is planned in the sense that d_i is the amount adjusted by the firm and to which costs are unambiguously linked. We maintain maximal discharge, $\bar{Z}_i$, as a deterministic magnitude. This assumption can, of course, be changed without involving any consequences for the qualitative

Figure 7.1: Damage and environmental indicator functions

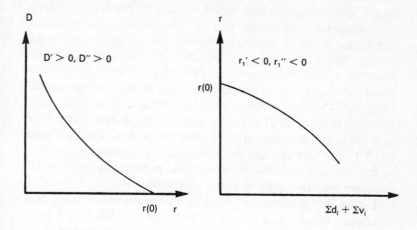

conclusions we shall draw. We are thus not specifically distinguishing between the two ways in which the discharge can be uncertain, that is uncertainty as to what is discharged from the firm and what reaches the recipient. If we stick to d_i as a realised magnitude, then uncertainty will only be due to transport from factory to recipient. If we consider d_i as a planned magnitude, or the magnitude the purification process should normally produce, and to which costs are linked independently of actual discharge, then v_i can be interpreted as the 'correction' of d_i, in order to give the actual discharge reaching the recipient.

We are now faced with the economic problem of how much each individual firm is to purify, if the objective is to minimise total purification and environmental costs. As we have now included stochastic variables, the expectation of total costs is minimised:

$$\min E\left\{ \sum_{i=1}^{n} c_{p,i}(\bar{Z}_i - d_i) + D(r(\sum_{i=1}^{n} d_i + \sum_{i=1}^{n} v_i, u))\right\} \quad (7.2)$$

Necessary first order conditions will then be:

$$c'_{pi} = E\{D'\}E\{r'_i\} + \text{covar}\{D', r'_i\} \quad (i=1,\ldots,n) \quad (7.3)$$

We shall here concentrate on interior solutions. (A discussion of corner solutions will be analogous with the discussion in 6.2.)

149

Operating with certain values for marginal damage, D', and marginal change in the environmental indicator r'_1, we get the well-known condition that sufficient must be purified to ensure that marginal purification cost $c'_{p,i}$ is equal to the marginal reduction in environmental services, calculated in terms of money. When uncertainty is included, another factor is introduced into this calculation, namely, the covariance between marginal damage and marginal change in the environmental indicator. The sign of the covariance will determine whether it would now be optimal to purify more or less in each factory, in relation to the situation without uncertainty. The righthand side in (7.3) does not depend on *which* factory we are considering, so that the condition for equal marginal purification costs for all factories will also apply in a state of uncertainty.

Let us first assume that there is no covariance between the stochastic variables, Σv_i, and u. With the conditions we have imposed on the functional forms in (7.1) and illustrated in Figure 1, the result we then get is that, for given total planned discharge, marginal environmental service production, r'_1, will be greater the smaller the realised value of the stochastic discharge component, Σv_i, is. The smaller Σv_i is, the greater D' will be. When Σv_i and u are independently distributed, we shall therefore get:

$$\text{covar}\{D', r'_i\} > 0$$

In every firm more must be purified when uncertainty is taken into account than in the event of certain expectations, when the purification cost function is convex ($c''_{p,i} > 0$), as assumed in section 6.2. The expected marginal environmental damage, calculated in terms of money, will have a risk supplement.

Positive covariance between Σv_i and u means that the realisation of an unfavourable discharge variation will coincide with an unfavourable situation for the environmental indicator function; the discharge will be greater than planned, Σd_i, while at the same time the extent of environmental services, R, is reduced for a given level of discharge. This constellation reinforces the positive covariance between marginal damage D' and 'marginal productivity' r' in marginal service production.

Negative covariance between the stochastic variables means that an unfavourable discharge situation will occur together with a favourable situation for the production of environmental services. At this stage we cannot say anything in general about the sign of the covariance without introducing specified functional forms and without doing empirical tests. It is interesting to note that the covariance *may* be negative. This means that less purification should be undertaken than in a situation involving certain expectations.

Covariation between the stochastic variables discharge $\Sigma\, v_i$ and the environmental service variable u might possibly arise as a result of background stochastic variables, such as weather conditions, wind, precipitation, sun, etc. An example of positive covariation could be the formation of photochemical smog. Favourable wind and radiation conditions reduce the amount of discharge reaching the recipient, while at the same time the formation of photochemical smog will be rendered more difficult. An example of negative covariation could be the purification of a sewage system which at the same time collects rain water. Purified sewage is discharged into a river. Violent downpours of rain may result in more organic discharge being carried past the sewage purification plant, while at the same time the recipient, the river, receives a greater supply of water, and is thus better able to break down the organic discharge.

In this social optimisation problem, we have not introduced any risk aversion. If the relevant authority considers this relevant, the risk aversion would have the effect of increasing the risk supplement to expected marginal environmental damage, calculated in terms of money.

In section 6.2 the environmental aspect was taken care of by setting a constraint on how low environmental service production could be. With stochastic discharge and environmental service function, the use of a constraint, instead of calculating environmental services in terms of money, can now be expressed by stating that the probability for going below a lower limit is to be below a given desired level, w: $U,\ P_r[R \leq \bar{R} = r(\Sigma\, d_i + \Sigma\, v_i, u)] < w$. (7.3) can now, as previously, be precisely what is to be the basis for imposing a charge. It would be unrealistic to suppose that it will be possible to measure the discharge that actually reaches the recipient. The basis for a charge can then either involve planned discharge or realised discharge, measured at the factory gate. In the latter case the factory will be faced with an uncertainty problem.

151

7.3 THE TAXMAN COMETH

7.3.1 Introduction

As a starting point we are assuming that the authorities have decided to impose effluent charges as a means of regulating the amount of pollution. The environmental protection authorities are in a position to measure conditions in the recipients involved, and arrive at a charge that will give the environment in the recipient a desired quality. This system presupposes that firms respect the rules and regulations for effluent charges. If no supervision of any kind is exercised, firms will be in a position to evade charges by discharging more effluent than their own charge returns would warrant. For this reason the environmental protection authorities, as well as the tax authorities, are anxious to supervise the sources of discharge.

In the next section we shall make use of the following simplified presupposition with regard to the institutional framework: the authorities are not in a position to undertake continuous measurements, either for economic or for technical reasons. We may note that as far as international experience of the use of effluent discharge is concerned, in none of the better-known cases do the authorities base themselves on continuous measurements of discharge. The charge has to be paid for a period taken as a whole.

The firm's discharge is not evenly distributed throughout the whole year. We might, for instance, imagine a pulp plant with production fluctuating during the course of the year owing to varying market conditions, on which the authorities lack perfect information.

The optimal amounts of discharge for the society are characterised by the marginal purification costs of discharge sources being equal to the marginal damage inflicted on the environment. The aim of the official charge policy is to get firms to adjust marginal purification costs to a charge rate for the whole year. The charge rate should as far as possible correspond to marginal damage in the course of the year.

We shall consider two approaches available to the authorities in effecting a solution to the charge: (a) the discharge amount for the entire year is estimated by making a random check on one of the days. The firm must then pay a charge equal to the result of the random check multiplied by 365 days; (b) firms themselves submit reports on their daily discharge, either implicitly in the payment of an effluent charge, or explicitly in a return of charges to the authorities. The number of effluent charge instalments in the course of a year will determine how often the authorities receive information from the firms on the amount of discharge. A random check is carried out on one day, and the result is then compared with the information the firm has submitted in its effluent charge return. If there is no agreement between the result of the discharge check and the information, a penalty is imposed on the firm.

The problem facing discharge source referred to as the firm is to minimise the total expected purification costs and payments of effluent charges, or to minimise the expected utility of these costs. As firms have no idea when a random check will take place or when one has been made, they will be compelled to make an adjustment under uncertainty. We shall draw attention to certain aspects of an adjustment of this kind undertaken under uncertainty. In particular we shall show that, where firms themselves submit reports, only in certain exceptional cases can one with the aid of a charge solution succeed in implementing a standard solution in which marginal purification costs are equal to marginal damage. The exceptional cases are characterised by the fact, *inter alia,* that firms are sufficiently unprincipled: they must cheat a little if an optimal allocation of resources is to be realised with the aid of charges.

7.3.2 Social optimum

We shall further simplify our analysis by presupposing that the amounts firms wish to discharge every day in the course of a year, without any effluent charge, are given and known magnitudes for each firm. These primary or maximal discharges may, for example, be unambiguous functions of production levels every day in the course of a year. It is sufficient for our purposes to consider the adjustment of one firm. The firm's purification cost function is as in Chapter 6, section 6.2.

$$c_i = c(\bar{z}_i - d_i); \quad c' > 0, c(0) = 0 \qquad (7.4)$$

where $\bar{z}_i$ is the given primary discharge in the sub-period i, and

153

d_i the actual discharge to the recipient. The actual purification cost function is the same for all sub-periods, while the given primary discharges and the actual discharges may vary from one sub-period to another.

The social problem of adjustment is to minimise the firm's total purification costs and the community's environmental damage in the course of the year.

$$\min_{d_1,\ldots,d_n} \left\{ \sum_{i=1}^{n} (m_i\, c\, (\bar{z}_i - d_i) + m_i\, D(d_i)) \right\} \tag{7.5}$$

$$\sum_{i=1}^{n} m_i = N \tag{7.5a}$$

Environmental damage is measured in a monetary unit by the function $D(d_i)$ which represents a simplification of the damage function (7.1). We have, furthermore, made it possible for the primary discharge level or output levels to be constant over a period of several days in the course of the year. The year is divided up into m_i periods with the same primary discharge within each period. N is the number of days in the year.

Necessary optimum conditions are:

$$c'(\bar{z}_i - d_i) = D'(d_i) \tag{7.6}$$

The marginal purification costs are therefore to be equal to the marginal damage for each period. Marginal damage may generally vary from one period to another. (In that case a subscript i should be added to the D-function.) Only if the damage function were to prove linear do we get the same marginal damage in each period. A common charge, t, in the course of the year can either be justified on this basis or because it has been institutionally decided that the charge must be the same for the whole year. In the latter case we can consider the following aspect of the problem: if by way of a digression we consider that the authorities are familiar with the primary discharges in each sub-period, the common rate of charge, t, can be determined by solving a so-called 'second-best' problem: (7.5) is minimised given the following side condition:

$$c'(\bar{z}_i - d_i) = t \tag{7.7}$$

154

The marginal purification costs for each period are to be equal and equal to the charge rate. The necessary first order conditions for this problem will be:

$$c'(\bar{z}_i - d_i) - D'(d_i) + \frac{\lambda_i}{m_i} c''(\bar{z}_i - d_i) = t \qquad (7.8)$$

$$\sum_{i=1}^{n} \lambda_i = 0 \qquad (7.9)$$

λ_i is a Lagrange coefficient associated with condition (7.7). Equation (7.8) may be used to study the social loss involved in having to introduce the practical arrangement of an effluent charge rate per discharged amount that remains unaltered throughout the year.

In continuing we shall assume a uniform charge for the whole year. For this reason (7.6) describes the social optimum. It is then immediately obvious from (7.6) and (7.7) that

$$d_j^{opt} - d_i^{opt} = \bar{z}_j - \bar{z}_i \qquad (7.10)$$

so that

$$\bar{z}_j > \bar{z}_i \rightarrow d_j^{opt} > d_i^{opt} \qquad (7.11)$$

'Opt' indicates the discharge amounts satisfying (7.6) and (7.7). If we consider two sub-periods j and i it is, in fact, optimal to discharge more in sub-period j than in i, if the primary discharge is greater in period j than in i; or, to put it differently, the difference between optimal discharges is to be equal to the difference between the primary discharges when we consider the periods pairwise.

7.3.3 Adjustment undertaken by the firm

(a) Introduction

The firm has no idea when the inspector will turn up, nor, in principle, when he has paid his visit.

The firm operates with subjective probabilities as to when in the year the inspector will turn up, that is, in which of the n subperiods he will arrive. What probabilities would it be reasonable to assume that the firm uses for the random check in each subperiod? If the firm believes that the check is carried out randomly, it is reasonable to calculate the probability for a visit, p_i, as

$$p_i = \frac{m_i}{N} \tag{7.12}$$

Are there any reasons why the firm should not use the probability structure (7.12)? One reason might be that the firm suspects that one or more periods are more likely than others. Another possibility is that the firm prefers to use a higher probability for visits in periods in which primary discharges are high. This can be interpreted as an expression of risk aversion. It might be natural to suppose that the environmental protection authorities will carry out several checks in the course of the year, partly in order to ensure improved supervision and partly to prevent firms making use of a probability structure that deviates from (7.12). The charge basis on which the authorities will operate will, on an average, comprise the amount discharged for the number of days on which checks are taken, multiplied by the total number of days. If the firm possesses information on the number of checks, for example, K, then it knows that there are

$$\binom{N}{K}$$

possible combinations.

If the environmental protection authorities suspect that probability structure (7.12) is not adhered to, a feasible strategy might then be to announce that they will be undertaking a certain number of checks. Since checks are secret, firms have no possible way of checking how many times they are carried out. If we calculate with a certain cost for the taking of checks, it might be optimal for the environmental protection authorities merely to carry out *one* check, but to announce that they will be carrying out several.

In the following we shall analyse only a one-visit variant. When the probability structure is as in (7.12) we shall call the probability structure correct.

(b) Basis of effluent charge determined by random check supervision

Risk neutrality. If the firm is 'risk-neutral' its adjustment problem involves minimising the sum of purification expenses and expected charge payment:

$$\min_{d_1,\ldots,d_n} \left\{ \sum_{i=1}^{n} m_i\, c(\bar{z}_i - d_i) + t \cdot N \sum_{i=1}^{n} d_i p_i \right\} \tag{7.13}$$

$$\sum_{i=1}^{n} p_i = 1 \tag{7.13a}$$

The necessary first order conditions for an internal solution will be:

$$c'(\bar{z}_i - d_i) = t \cdot p_i \frac{N}{m_i} \tag{7.14}$$

As long as the firm uses a correct probability structure (see (7.12)), condition (7.14) will be reduced to a situation in which marginal purification costs will be equal in all periods and equal to the rate of effluent charge per discharge unit per day. Thus if the firm chooses the correct probability structure (7.12) then the social optimal solution (7.6) will be realised. If the firm's subjective probabilities differ from (7.12) we get a social wrong adjustment. If the probability of receiving a visit during the sub-period is estimated as lower than that given in (7.12), then (7.14) immediately indicates that the marginal costs will be lower than in the optimal solution (7.6). Since the primary discharge is given, a lower marginal purification cost means that the discharge d_i must be greater than what is socially optimal. Correspondingly, when the probability p_i is set at a higher level than in accordance with rule (7.12) the result we get is that the discharge will be less than the correct discharge in accordance with solution (7.6).

Risk aversion. If the firm desires to use a higher probability for random checks than that given in (7.12) in a period with high primary discharge, then this can be interpreted to mean that the firm has risk aversion. An alternative way of dealing with this is

157

to presuppose that the firm compares the various possibilities for discharge with total costs, that is, the sum of purification costs and the payment of effluent charge, by allocating a preference function to costs:

$$- U \left(\sum_{i=1}^{n} m_i \, c(\bar{z}_i - d_i) + t \cdot N \, d_j \right), \tag{7.15}$$

where d_j is the discharge measured by the environmental protection authorities when making a check. The function $- U$ is concave, so that U is convex; the utility diminishes, the greater total costs become. We shall call y_i total costs when d_j represents the results of the random check undertaken by the environmental protection authorities. The relationship between two total costs with respectively samples d_j and d_i will then be:

$$y_j = y_i + t \cdot N \, (d_j - d_i) \tag{7.16}$$

For every possible d_j the firm in turn allots a probability p_j. The firm is assumed to maximise expected total utility:

$$\max \left(- \sum_{j=1}^{n} p_j \, U(y_j) \right) = \min \sum_{j=1}^{n} p_j \, U \left(\sum_{i=1}^{n} m_i \, c(\bar{z}_i - d_i) \right.$$

$$\left. + t \cdot N \, d_j \right) \tag{7.17}$$

The necessary first order conditions are:

$$\sum_{j=1}^{n} p_j \, U'(y_j)(- m_i c'(\bar{z}_i - d_i)) + p_i \, U'(y_i) \, t \cdot N = 0 \tag{7.18}$$

In order to facilitate comparison with (7.6) the conditions can be transformed to:

$$c'(\bar{z}_i - d_i) = t \cdot p_i \frac{N}{m_i} \cdot \frac{U'(y_i)}{\sum_{j=1}^{n} p_j U'(y_j)} \tag{7.19}$$

Let us now assume that the firm presupposes that a check will be carried out at random, that is, that (7.12) applies. We can then see from (7.19) that even if the probability structure is correct, condition (7.6) will no longer be generally fulfilled. The last term in (7.19) expresses the ratio of the marginal disutility of total costs, when d_i is the basis for calculating the effluent charge, and expected marginal disutility. With probability structure (7.12) average marginal disutility is used as an estimate for expected marginal disutility. With a correct probability structure (7.12) the result therefore is that if marginal disutility, with the discharge of period i as a sample, is lower than average marginal disutility, then the righthand side of (7.19) is lower than t. This results, furthermore, in discharge being greater than optimal. Generally speaking we get:

$$U'(y_i) \underset{<}{\overset{\geq}{=}} \frac{1}{N} \sum_{j=1}^{n} m_j U'(y_j) \Rightarrow d_i \underset{>}{\overset{<}{=}} d_i^{\text{opt}} \tag{7.20}$$

Since total purification costs are constant, and the same for all values of y_i, then y_i will be smaller, the smaller d_i is. Furthermore, U is convex, so that a low marginal disutility is associated with a low y_i. We can then draw the following conclusion: risk aversion results in small discharges becoming too great, and large discharges being too small, in relation to an optimal solution characterised by (7.6). Only in a case with linear utility function and correct probabilities (7.12) is the firm's adjustment in agreement with (7.6). A linear utility function means the same as risk neutrality, that is, we are back to the case dealt with previously.

(c) Independent reports and random check supervision

The institutional system is now such that the firms themselves send in reports on the basis for the effluent charge. By way of control the authorities make random checks in the course of the year. If it turns out that the firm has reported a lower discharge on the actual day than the check shows, then the firm is fined. The authorities calculate that the difference between actual discharge and reported discharge on the day when the check is made applies to every day. For this extra discharge a penalty rate of s per discharge unit per day has to be paid.

Firms pay an effluent charge of t per kg per day on the dis-

159

charge which they have reported in the effluent charge returns. If we let x_i indicate the discharge amount they operate with in their returns to the authorities, then the firm's stated payment of effluent charge in the course of the year will be

$$t \sum_{i=1}^{n} m_i x_i$$

The actual discharge on one day in period i when the check is carried out, is d_i. Clearly we shall get

$$x_i \leqq d_i$$

The authorities compare the result of their check d_i with the reports made by the firm, and which are implicitly given in the effluent charge payment. In cases of discrepancy a penalty tax is imposed on the firm.

How often payment takes place and/or when returns are submitted to the authorities will determine what possibility for comparison is available to the authorities. We consider two alternatives:

Alternative 1: d_i in relation to x_i

Alternative 2: d_i in relation to $\bar{x}$ where $\bar{x} = \dfrac{1}{N} \sum_{i=1}^{n} m_i x_i$.

Alternative 1 presupposes that the authorities receive frequent payments from the firm in the course of the year. Alternative 2 presupposes *one* payment in the course of the year. By way of comparison, in the Norwegian system of VAT, payments are made in six instalments. However, we shall assume that payment of charges is so frequent that a comparison between the random check results, d_i, and the amount given as discharge, x_i, may be undertaken on the basis of Alternative 1. The case of less frequent payments, with Alternative 2 as a limiting case, is easy to analyse.

As already mentioned, the authorities either ignore or are unaware of seasonal variations in the firm's production. Consequently, should the authorities discover that the sample result d_i differs from the amount given, x_i, they impose a penalty charge

on the firm such that the annual basis for this surplus charge
equals the observed difference multiplied by the number of days
in the year. If we call the penalty charge per kilo per day s, the
amount the firm will have to pay by way of penalty on an
annual basis, if it is caught cheating, will be $sN(d_j - x_j)$.

Risk neutrality. The firm's adjustment problem is now
described as follows:

$$\min_{x_i, d_i (i=1, \ldots, n)} \left\{ \sum_{i=1}^{n} m_i \, c(\bar{z}_i - d_i) + t \sum_{i=1}^{n} m_i x_i \right.$$

$$\left. + sN \sum_{i=1}^{n} (d_i - x_i) \, p_i \right\}, \qquad (7.21)$$

$$\sum_{i=1}^{n} p_i = 1 \qquad (7.22a)$$

$$0 \leq x_i \leq d_i \leq \bar{z}_i \qquad (7.22b)$$

x_i is the reported discharge amount per day, and s is the rate of
fine per unit of discharge per day. The firm must decide what
amounts of the polluting substance are actually to be discharged
and what discharged amounts are to be reported to the authorities. In other words we are assuming that the firm is considering whether it is worth while cheating. Necessary first order conditions for adjustment of actually discharged amounts will be:

$$c'(\bar{z}_i - d_i) = s \, p_i \frac{N}{m_i} \qquad (7.23)$$

If the firm now has the correct probability structure (7.12) we
can observe that the actual discharges are only socially correct if
the rate of penalty imposed s is equal to the effluent charge rate
t. If the penalty charge is greater, the actual discharge will be too
small in each period in relation to optimal discharges.

With regard to determination of the reported amount of discharge, the necessary first order conditions give us the following:

$$\frac{t}{s} \; {\geq \atop <} \; \{ p \frac{N}{{}^i m_i} \} \quad \text{gives} \quad x_i \quad \left\{ \begin{array}{l} = 0 \\ {}[0, d_i] \\ = d_i \end{array} \right. \tag{7.24}$$

(7.23) and (7.24) now give us the surprising conclusion that with the interior solution of (7.24), (7.23) and (7.24) will give us the condition for the social optimum (7.6). If the probability structure is correct, in the sense that it corresponds to (7.12) an interior solution demands that the penalty charge should be equal to the ordinary charge. In a situation of this kind the allocation of resources is optimal, since (7.6) is satisfied. As there has to be an interior solution of (7.24), however, firms will have to cheat a little, and be sufficiently unprincipled for this optimal allocation of resources to be optimal. If condition (7.6) is to be realised, the authorities will, in fact, have to shut their eyes to an evasion of charges, in the sense that evasion is not punished more severely than the reported discharge. In cases where firms operate with 'wrong' probability structures, this last-mentioned conclusion will have to be altered.

When the tax rates s and t and/or the probability structure are such that the firm chooses to report correctly $(x_i = d_i)$ or nothing $(x_i = 0)$, we can see from (7.23) that the allocation of resources no longer satisfies the optimal condition (7.6). When the firm chooses to play the complete villain, we can see from (7.23) and (7.24) that too *little* is purified in each period.

When the firm chooses to be honest, by making complete returns, we can see from (7.23) and (7.24) that too *much* purification is carried out in each period. A strategy of honesty being the best policy also provides a cleaner environment. The point is that it would be far too pure, in relation to the social optimum. In other words, an optimally purified society consists of moderate villains.

From (7.23) we can see that when the probability structure is wrong, and of such a nature that the firm overestimates the probability of a visit in periods in which primary discharges are greatest, too much purification will be undertaken during these periods. Correspondingly, too little purification will be carried out in periods in which primary discharges are smallest. The big discharges will thus be too small and the small discharges too big.

Risk aversion. If a firm wishes to use a higher probability for random checks than that given in (7.12) during a period with high primary discharge, this may be interpreted as indicating that the firm has a risk aversion. An alternative method of dealing with this is, as above, to presuppose that the firm compares with various possibilities for the outcome of total costs, namely, the sum of purification costs and the payment of effluent charges, by evaluating costs through a preference function:

$$-U(y_j)$$

where

$$y_j = \sum_{i=1}^{n} m_i \, c(\bar{z}_i - d_i) + t \sum_{i=1}^{n} m_i x_i + sN(d_j - x_j)$$

d_j is the discharge which the environmental protection authorities measure when taking a random check. The function $-U$ is, as above, concave, so that U is convex, and the utility is less, the greater total costs are. We call y_j the total costs when d_j represents the environmental protection authorities' sample. The relationship between two total costs with samples d_j and d_i respectively will now be:

$$y_j = y_i + sN[(d_j - x_j) - (d_i - x_i)]$$

For every possible outcome of the random check the firm establishes a probability p_j. The firm is assumed to maximise the expected total utility:

$$\max_{d_i, x_i} - \sum_{j=1}^{n} p_j \, U(y_j) = \min_{d_i, x_i} \sum_{j=1}^{n} p_j \, U(y_j)$$

(7.22a-b) also applies to this problem.
   The first order conditions are:

$$c'(\bar{z}_i - d_i) = s \, p_i \frac{N}{m_i} \cdot \frac{U'(y_i)}{\sum_{j=1}^{n} p_j U'(y_j)} \qquad i = 1, \ldots, n \qquad (7.25)$$

163

$$\frac{U'(y_i)}{\sum\limits_{j=1}^{n} p_j U'(y_j)} \begin{array}{c} < \\ = \\ > \end{array} \left\{ \begin{array}{c} t \, m_i \\ \overline{} \\ s \, p_i N \end{array} \right\} \Rightarrow x_i \left\{ \begin{array}{l} = 0 \\ [0, d_i] \;\; i = 1, \ldots, n \\ = d_i \end{array} \right. \quad (7.26)$$

In the previous section a correct probability structure, namely (7.12) and $s = t$, produced the result that the social optimum condition was satisfied. This is no longer the case. We now see from (7.25) that just as in (7.20) we shall have

$$U'(y_i) \begin{array}{c} > \\ = \\ < \end{array} \frac{1}{N} \sum\limits_{j=1}^{n} m_i \, U'(y_j) \Rightarrow d_i \begin{array}{c} < \\ = \\ > \end{array} d_i^{\text{opt}}$$

This means that when the marginal disutility in a period i is greater than the average marginal disutility, the firm's discharge is less than the social optimum. From the definition of y_i we see that y_i increases when d_i increases. We consequently get as an answer, just as above, that the risk aversion results in big discharges being too small and small discharges being too big. From (7.26) we can see that there will be less cheating when the maximum discharge is greatest.

In this case, too, the optimum condition (7.6) is satisfied by interior solution in (7.26). When there is a little, but not total, cheating, we can therefore have the social optimal amount of pollution implemented. If firms in addition have a correct probability structure, we can see from (7.26) that all the y_i's would then have to be equal. From the definition of y_i we note that this means that all $(d_i - x_i)$ must be equal. Given a correct probability structure and given that we are to have the social optimum described in (7.6) realised, we must therefore allow for a certain amount of cheating in every period. At the same time a solution must exist that makes the cheating in kilos equal from period to period. Since all the y_i's must be equal in a situation of this kind, the marginal disutility in a period i will be equal to the annual average. From (7.25) it then follows that this solution demands that $s = t$. Thus evaded discharge amounts must be penalised on a par with amounts that have been reported.

With a 'wrong' probability structure the optimum condition (7.6) can be satisfied without $(d_i - x_i)$ being equal from one period to another.

Irrespective of what the probability structure is, the result we

164

shall get, however, as in the previous section, is that fulfilment of the social optimum conditions (7.6) requires that the players should be suitably unprincipled or villainous. And with a correct probability structure the authorities should turn a blind eye to cheating. The penalty charge should be fixed at a level equal to that of the ordinary effluent charge.

In cases of honesty ($x_i = d_i$) we see from (7.26) that

$$t < s p_i \frac{N}{m_i} \frac{U'(y_i)}{\sum\limits_{j=1}^{n} p_j U'(y_j)}$$

From (7.25) and (7.6) we can see that now, too, with honesty accepted as the only true policy, too much purification will be carried out in all periods. The community, in fact, will be *too* pure.

Conversely, we have no difficulty in observing that in a community that is thoroughly unprincipled there will be too little purification in all periods. The community will as a result be *too* polluted. The point once again is that one should aim at a suitable degree of pollution which also demands a modicum of villainy.

SELECTED REFERENCES AND FURTHER READING

Allingham, M. and Sandmo, A. (1973): 'Income tax evasion', *Journal of Public Economics*, pp. 323-38
Baumol, W. and Oates, W.E. (1975): *The theory of environmental policy*, Prentice Hall, NJ
Just, R.E. and Zilberman, D. (1979): 'Asymmetry of taxes and subsidies in regulating stochastic mishap', *Quarterly Journal of Economics*, pp. 139-48
Mäler, K.G. (1974): *Environmental economics: a theoretical inquiry*, The Johns Hopkins Press, Baltimore

8

Long-term Problems in Connection with Pollution

8.1 POLLUTION CAUSED BY CURRENT DISCHARGE AND DISCHARGE ACCUMULATION

We can distinguish between two types of pollutant discharge:

(a) Pollution due to current discharge of residuals;
(b) Pollution due to residuals discharged in the past.

The first kind may be said to be due to day-to-day flow of residuals. This might, for example, be air pollution, where the damage this type of air pollution inflicts on the individual at a given point of time is independent of how much of this air pollution previously existed. This form of pollution has been the object of the greatest concern in public debate. When pollution is due to the accumulation of residuals discharged at a previous point of time, it becomes particularly necessary to study pollution on a long-term basis. Today's discharges will then produce negative effects in the future.

In order to concretise pollutant type (b), we might consider a few examples. This might involve discharges of organic materials in water from the pulp and paper industry, where continuous discharges are not broken down so rapidly by natural processes that gradual reduction of the oxygen in the water can be prevented. The self-purification capacity of the water will depend on natural conditions, but also on how greatly the water has previously been contaminated. A lake or a watercourse will not suddenly be incapable of any other use than as a deposit for waste. The process takes time. Another example might be the discharge of heavy metals, such as mercury and cadmium, or a

166

substance such as PCB, or pesticides, such as DDT. A common feature of these substances is that either they are not broken down by natural processes or they are broken down very slowly. The point then is that any danger resulting to individuals will not as a rule be due to acute poisoning, but the result of repeated and possibly small doses. A gradual concentration takes place, *inter alia*, through food chains. Thus, for example, mercury may be constantly discharged into drinking water, without damage arising as a result of drinking this water, but eating fish caught in this water may have injurious effects.

In the case of air pollution, too, pollutant type (b) may occur. Discharges from motorised traffic and industrial firms may cause health injuries not due to immediate poisoning, but as a result of the intake of small, but repeated, doses. Discharges of gas containing fluorine and particles from aluminium works may result in necrosis of the jaw. There are numerous examples of present-day pollution being due to previous discharges. The accumulation phenomenon is not an exception, but rather the rule in pollution problems.

The table below gives figures for discharges in Norway in 1970 of three residuals. These three residuals all possess the property of accumulation in the natural environment. With the aid of a multi-sectoral growth model, cf. Førsund and Strøm (1974), we have calculated discharges every year up to the year 2000. These discharge figures are not given here, but we present figures for amounts contained in the natural environment of these three substances for the year 2000. Stocks are calculated by adding together all discharges from 1970 to 2000. Owing to insufficient data we have put stocks for the initial year (1970) as equal to 0. Calculations take into account changes in the industrial structure, but do not take into account future technical changes in the individual firms, for example, those associated with purification, recycling, etc. as a result of more restrictive environmental policy. The import of substances from other countries and the export of substance to other countries, too, have not been taken into account.

The calculations in Table 8.1 thus present a more horrifying picture ('What will happen if nothing is done?') than a prognosis of what will actually occur.

Calculations show that if nothing is done with discharges of the accumulated substances at an early stage in what is left of this century, by the turn of the century considerable amounts of

167

Table 8.1: Discharge of three accumulated substances, as well as deposits[a] in the natural environment in the year 2000

Source of discharges	Lead		Mercury		Pesticides	
	1970 Discharges (tonnes p.a.)	2000 Stocks (tonnes)	1970 Discharges (tonnes p.a.)	2000 Stocks (tonnes)	1970 Discharges (tonnes p.a.)	2000 Stocks (tonnes)
Agriculture	6	207	0.9	31.0	1300	44,922
Forestry	1	63	0.1	6.0	200	12,526
Mining	1	57	—	—	—	—
Fertiliser industry	—	—	2.0	120.0	—	—
Primary iron and metals	1,700	81,286	1.0	48.0	—	—
Electrotechnical industry	5	360	—	—	—	—
Other industry	3	216	—	—	—	—
Transport[b]	78	5,042	—	—	—	—
Defence	5	229	—	—	—	—
Households	438	21,812	—	—	—	—
Total	2,237	109,272	4.0	205.0	1500	57,448

Notes

a. Stocks in the initial year 1970 have been put at zero. The reason why 1970 is chosen as the initial year is that for this year a considerable effort was undertaken to estimate a wide variety of discharges, cf. Førsund and Strøm (1974).

b. The use of private cars has been included under households.

polluted substances will have been spread around in the natural environment, involving great potential injurious effects. In the course of these 30 years it will be noted that very substantial amounts of lead and pesticides will be accumulated, as well as considerable amounts of mercury. It will be seen that growth cannot be linear, since 30 times the 1970 discharge results in far smaller deposits than those calculated. Pollutant deposits grow more rapidly than this. Deposits of pesticides increase somewhat more slowly than the two other deposits of polutants. This is due to the fact that pesticide discharges emanante from two sectors (agriculture and forestry) which expand at a slower rate than, for example, the manufacturing sectors.

8.2 WHAT IS THE SIGNIFICANCE OF THE ACCUMULATION PHENOMENA?

If we consider the substances mentioned in the previous section, we might be tempted to raise the following question: could not the use of these substances be prohibited? As a rule the situation is not quite as simple as this. The point is that the use of these substances offers immediate advantages. Thus, taking the short-term view, their use has a favourable effect on the economy. Negative effects are long term and make their presence felt via accumulations in the environment. In the case of pollution type (b), as defined in the previous section, the problem therefore arises: to what extent should we reduce discharges of these substances today in order, in the future, to ensure less pollution? In making assessments of this kind, in other words, one is considering the environmental capacities of the future. This is a similar problem to that for traditional economic growth: how much should we reduce consumption today in order to build up real capital so that consumption may increase in the future?

To what extent society would be willing to make sacrifices now, in order to avoid damage later on, will depend on the individuals' preferences in the community or, if such decisions are centralised, the preferences of the authorities. In such long-term questions the attempt must be made to answer the following questions:

What importance is attached to the future as opposed to the present?
What environmental stocks do we want the next generation to 'inherit'?

Pollution of type (b) will, of course, also be capable of damaging the present generation.

We must, in fact, try to decide what we want a future planning period to inherit by way of stocks either in the form of real capital, human capital, or environmental capital. Composition, too, must also be taken into account. It is realistic to assume that an increased inheritance of real capital must take place at the expense of the natural environmental capital. Preferences will probably be attached to the scope of choice with which the following generation is to be presented. It is by no means certain that the greatest liberty of choice will be available by concentrating solely on making natural capital as great as possible. An assessment must be undertaken, both with regard to the size and the composition of the inheritance. What is important is therefore: If pollution is associated with current discharges of

169

residuals a 'generation' can at a given point of time itself decide the amount of pollution merely by passing legislation on current production and current consumption. Previous history will then have no bearing on pollution at a given point of time. The effect that prehistory in a situation of this kind may have involves the composition of, *inter alia*, real capital equipment in the community. Compositions may be wrong, *inter alia* on account of insufficient control of the market economy. This type of pollution, however, will in any circumstances impose few obligations on the choice to be made by future generations.

If pollution is due to the second type, that is, that the accumulation aspects are essential, the situation will be different. Pollution will then also have to be associated with previous discharges of residuals, and not only with current discharges. Reducing discharge 'today' will not only have a bearing on pollution 'today'. A generation must accept pollution as a heritage handed down by previous generations.

The dynamic processes in the natural environment when residuals are discharged may cause irreversible processes in the environment. In the first place this may be the case owing to the fact that substances are discharged which assimilate in the natural environment, and are only slowly broken down. In the second place threshold values can be exceeded after a certain accumulation state in the natural environment, so that major ecological changes may occur.

In many cases it may prove difficult to predict the long-term effects of discharge flows on the natural environment. Random circumstances may have a major effect, influencing the self-purification capacity of the natural environment. This uncertainty with regard to the effects of discharge flows, the possibility of irreversibility, etc. should encourage public authorities to exercise caution in their long-term policy, cf. Chapter 9.

8.3 MODEL FOR ANALYSING DYNAMIC POLLUTION PROBLEMS

8.3.1 Introduction

Let us imagine a society where a good is produced that consumers in that society benefit from. In the production of this good a quantity of a residual is generated proportional to the amount produced at any point of time. We assume that it is the sum of previous discharges of residual that at a given point of time causes pollution.

To simplify the problem we shall therefore presuppose that the amounts of residual discharged into the environment merely accumulate. There are no forces present to break down, thin out or in any other way reduce these or the pollutant effect these discharges have. This assumption will be relaxed in the mathematical appendix to this section.

The damage pollution inflicts on consumers can therefore be interpreted as depending on the stock of residuals present in the environment, not on the discharge at a specific point of time. The discharge is then 'stored' in the environment, and is added to the deposits of residuals already there from before. At a definite and given point of time (now) it is impossible to restrict the damage by controlling current discharge. Future damage, however, can be affected by changing the amount of discharge now. Is this in the interests of the community? This will, *inter alia*, depend on what importance the community *now* attaches to living conditions in the future.

What living conditions will be in the future is a matter of importance to most people. Nevertheless, there may be differences of degree. Intuitively, it seems obvious that the less importance one attaches to the future, the less one is willing to sacrifice anything now in order to enjoy a better life later on. In our problem: one is less willing to reduce the total consumption of the good which produces pollution in the future. The circumstance is central to an analysis of the problem of pollution we are going to study. By way of simplification, however, we shall disregard some of this and presuppose that equally great importance will be attached to the utility flows every year in the future. This narrower aspect of the problem will be relaxed in the appendix to this section.

8.3.2 The objective function

The consumer side will be dealt with summarily, two goods are specified, a consumer good and a pollutant. There is L_0 identical, individual and additive utility functions. The marginal utility of the consumer good is positive and decreasing with

171

increasing consumption. The marginal damage to each individual is constant. The instantaneous, social welfare function is simply the sum of the L_0 identical utility functions. The number of consumers, L_0, is constant over time.

Thus, the utility side of the problem can be specified in the following way. Let

X_t = total consumption at time t
U = individual utility
Z_t = stock of pollutants or residuals accumulated in the environment before t
g = the numerical value of the (constant) individual marginal damage of the stock of pollutants.

The social welfare function is thus (the subscript indicating time function is suppressed below where otherwise clear):

$$W = L_0 U \left(\frac{X}{L_0} \right) - L_0 g Z \qquad (8.1)$$

$$U' > 0, \; U'(0) = + \infty, \; U'' < 0, \; g > 0$$

The production side will also be very summarily dealt with. What does it cost firms to produce the amount of goods? In this connection costs mean the use of resources that could otherwise have been used for the production of other goods and services. We shall assume that costs only depend on the amount produced, and that these costs increase when the amount of production increases, and that the marginal cost rises with increase in production. We are assuming that all firms have similar costs functions.

If we call costs, c, then we get

$$c = c(X) \qquad (8.2)$$

and c' is marginal costs.

The current discharge of residuals, $\dot{Z}$, is proportional to production:

$$\dot{Z} = a \frac{X}{L_0} \qquad (8.3)$$

where a is a given constant.

If we subtract production costs from social welfare, W, we get the *current* social surplus resulting in the production of the quantity of goods X. If we call this social surplus π_t, we get

$$\pi_t = L_0 U\left(\frac{X_t}{L_0}\right) - L_0 \cdot g \cdot Z_t - c(X_t) \qquad (8.4)$$

It is assumed, in other words, that utility can be expressed in the same units as costs and damage. This assumption is highly restrictive: in the first place it presupposes that expenses for the good concerned comprise a small proportion of the consumers' total budget, that the good considered is independent of the need for other goods and services, that the prices of other goods and services are constant, and that the marginal utility of these other goods and services is constant. In addition, we must assume that the prices of all other goods and services are constant over time, that the amount of resources available for production of goods and services in the economy in terms of the prices of other goods and services remains unchanged over time, and that the population is stationary during this period. The model applies to a stationary economy, except for conditions in the market for the good under consideration.

The qualitative conclusions that will subsequently be introduced into this chapter will, however, not be changed should a more general framework be taken as a basis for analysis.

Since we are discussing pollution resulting from accumulation of the discharge of previous years, the objective function in (8.4) will not be sufficient. We can define the *long-term* social surplus as:

$$J_0 = \int_0^T \pi_t \cdot dt \qquad (8.5)$$

We have here 'added up' all future current surpluses from today (year 0) and up to the planning horizon T. By way of simplification we have allowed current surpluses to count equally, irrespective of where in the planning period they occur. This is a special assumption, since most people would probably rather accept an objective function, where the events in the near future

173

count more than events in the distant future. A more satis-factory objective function of this kind will be discussed in the mathematical appendix to this section.

8.3.3 The planning problem

Let us assume that a centralised decision is to determine how much is to be produced in order to make the long-term eco-nomic surplus as big as possible. By this is meant not a decision with regard to a particular quantity of production, but a con-nected series of production quantities. The collective body is to decide now (today) how much is to be produced now and every subsequent year until the termination of the planning period. In order that this planning is to have some meaning, the collective body must decide at the point of time of planning what amount of pollution any subsequent planning period is to inherit. In so far as we have presupposed that the good generating pollution is to be produced at every point of time, and in so far as we have presupposed that the growth of pollution is proportional to pro-duction, then the inheritance in this case will comprise an amount of pollution that is greater than that inherited by the present planning period. The decision on inheritance could now be, for example, that the amount of pollution that the next planning period takes over is to be less than or equal to a certain figure. It is in the inheritance decision now that the conditions of life in the next planning period will be taken into consider-ation. The type of inheritance decisions we have mentioned are only intended as examples. In order to make the following presentation simpler, we shall presuppose that the decision on inheritance is not particularly generous, namely, the amount of pollution to be inherited by the next planning period may be just whatever it likes to be.

The problem can be expressed more precisely in the follow-ing mathematical terms:

$$\max J_0 = \int_0^T \pi_t \, dt$$

given

$$\pi_t = L_0 U(\frac{X_t}{L_0}) - L_0 g Z_t - c(X_t) \qquad (8.6)$$

$$\dot{Z}_t = a X_t \qquad (8.7)$$

$$X_t \geqq 0 \qquad (8.8)$$

$$Z(0) = Z_0 \qquad (8.9)$$

$$Z(T) \text{ is free} \qquad (8.10)$$

Since L_0 is constant over time it can be set equal to 1. Eq. (8.7) gives the accumulation of residuals, and the stock of residuals produces pollution. aX_t is the discharge of residuals at the point of time t. a is the discharge per unit produced. This is assumed to be constant over time. For the moment it is assumed that no activities exist in the environment or the economy which break down or depreciate residual discharges. Eq. (8.8) indicates the constraint of non-negative total production. Eq. (8.9) indicates the initial stock of discharges. Eq. (8.10) indicates the inheritance decision. T is the given planning horizon.

The terminal condition (8.10) is not particularly generous to the next planning period. The specification has been selected since it simplifies subsequent calculation. Other but more complicated terminal conditions might arise, for example:

$$Z(T) \leq Z_T (= \text{a number})$$

From (8.7) and (8.8) we note that $Z(T)$, in any circumstance, must be greater than or equal to Z_0. It further follows that according to our specification of the problem it is not necessary to introduce the logically necessary constraint on the variable Z: $Z(t) \geq 0$ for all t. If Z_0 is not negative, $Z(t)$ will not be negative either for any $t > 0$. If the problem is to be meaningful, we must have $Z_0 \geq 0$.

The variable X will subsequently be called the control variable and Z the state variable. The control variable X governs the development of Z via the differential equation (8.7). The problem tells us that the collective decision-making body must choose a time function $X(t)$ (a 'connected series of production quantities') that maximises the objective function, given the conditions (8.7)–(8.10). The problem can be solved

175

with the help of the theory of optimal control as demonstrated in the appendix of this chapter.

8.3.4 The optimal solution

Instead of solving this problem on strictly mathematical lines, we shall in this section use a more intuitive method. Let us assume that there exists a connected series of production quantities of such a kind as to make the long-term economic surplus as large as possible, and where this series of quantities agrees both with what the present planning period has inherited in the way of pollution and what it has decided that the next planning period is to inherit. What characterises this optimal series of quantities?

We shall take our starting point in this series of quantities, and select a small interval of time and increase the production of the good in this interval by a small amount Δ. The community's utility of this good will then momentarily acquire an increment, namely:

$$MU_t \cdot \Delta \tag{8.11}$$

and otherwise no more. The marginal utility, $MU_t = U'(X_t)$, and marginal cost, $MC_t = c'(X_t)$, are the ones that apply in this small interval of time, but prior to the addition of Δ. The footnote t indicates that they refer to a period of time, for example, the beginning of the interval.

Variable costs will acquire an increment

$$MC_t \cdot \Delta \tag{8.12}$$

and otherwise no more.

Due to the fact that we have called the discharge of residuals per unit produced at any point of time for the constant a, the stock of residuals at each of the subsequent points of time will be

$$a \cdot \Delta$$

units greater. Due to the fact that we have called the marginal damage that the accumulation of residuals generates at every point of time the constant g, the damage will receive an increment equal to

$$a \cdot \Delta \cdot g(T-t), \tag{8.13}$$

or, if we call the increment in damage resulting from increased production by one unit at the point of time t, MD_t:

$$MD_t \cdot \Delta \tag{8.14}$$

where

$$MD_t = a \cdot g \cdot (T-t), \tag{8.15}$$

and where t is the beginning of the infinitely small interval of time and T is the length of the planning period. $T - t$ is therefore equal to the remaining part of the planning period. It is throughout the remaining period that the extra accumulation of residuals ($a\Delta$) is allowed to work, producing marginal damage equal to g at any point of time. Hence (8.13).

If the series of production quantities we have selected is to be the one that makes the economic surplus the greatest possible, the additional utility must be equal to the increase in disutility, that is:

$$MU_t \cdot \Delta = MC_t \cdot \Delta + MD_t \cdot \Delta \tag{8.16}$$

where $MC_t + MD_t$ is the total marginal disutility associated with production at the point of time t. Here, Δ, the increase in production, can be eliminated, and what we then get is that a necessary characteristic of a continuous series of quantities, making the economic surplus as big as possible, must at every point of time t in the planning period be:

$$MU_t = MC_t + MD_t \tag{8.17}$$

The problem discussed in this section is an example of one in the classic calculus of variation in mathematics. If we differentiate (8.17) with respect to t we get one of the necessary conditions to ensure that the 'continuous series of production quantities ensures the greatest possible economic surplus'. We have carried out this differentiation in (8.18). The procedure,

177

however, is highly informal. Eq. (8.18) corresponds to the Euler equation in a more strictly mathematical formulation.

If we simplify still further, and assume that marginal costs are constant, namely, of the same magnitude irrespective of how much is produced, then it follows from (8.17), which, of course, applies to every point of time in the planning period, that

$$MU_t - MU_{t-1} = MD_t - MD_{t-1}$$

Since

$$MU_t = MU(X_t) = U'(X_t)$$

then

$$MU(X_t) - MU(X_{t-1}) = -ag < 0 \qquad (8.18)$$

Thus,

$$MU(X_t) < MU(X_{t-1}) \qquad (8.19)$$

As assumed in (8.1) marginal utility decreases with increasing consumption. It then follows from (8.19) that $X_t > X_{t-1}$. Total consumption, which is naturally equal to total produced quantity, has therefore to grow over time. (8.19) applies to every point of time in the planning period.

The conclusion will therefore be: the continuous series of production quantities which makes the long-term economic surplus greatest is, given our assumptions, characterised by the fact that more and more is produced of the good producing pollution. This will also apply if MC increases with the level of production. This may seem paradoxical, but it is not. Note that so far we have said nothing about the level of production, for example, at the beginning or end of the planning period. Nor have we said anything about the relationship of these levels to production and the consumption that will be realised in a decentralised economy and where market participants are not led to pay sufficient attention to the negative side-effects of total consumption (or of the total production) in their assessments.

8.3.5 Solution based on perfect competition compared with the optimum solution

An economy based on perfect competition is a decentralised economy, where consumers maximise utility, given budget conditions, and producers maximise profit. None of these have sufficient incentive to consider that discharge 'today' would inflect damage on them 'tomorrow'. Adjustment on the part of the consumers results in the well-known phenomenon that the price is equal to marginal utility (according to our definition and measurement of utility), and in the case of producers the result is that the price is equal to marginal costs. Equilibrium will then be characterised by:

$$MU(X_t) = MC(X_t) \qquad (8.20)$$

In so far as we have implicitly assumed above that no shift over a period of time occurs either in utility or cost functions, at every point of time an equally large quantity of the good will be produced and consumed in the decentralised free competition economy.

If we compare (8.20) with (8.17) and (8.15), we observe that (8.17) coincides with (8.20) only at the point of time when the planning period has terminated, namely, when $t = T$. Thus, the centralised decision which takes into account all the disutilities involved gives an equally large production quantity *at this point of time* and therefore discharge of residuals as the unorganised mass of consumers and producers arrive at on their own account *at any given point of time*. This reflects the fact that at the terminal point of time no other disutility *per se* is involved than production costs in the production of the good. The damage for the disutility that pollution now involves has been taken into account in the 'ungenerous' inheritance decision. But what bearing does the centralised decision have on the size of production at the point of time prior to T? From (8.17) and (8.15) we obtain the following:

$$MU(X_t) - MC(X_t) = ag(T - t) \quad \text{for each } t < T. \qquad (8.21)$$

In other words, $MU - MC$ is positive. If we presuppose as above that MU decreases with an increasing consumption while

179

MC increases with increasing production, total consumption (= total production) X_t must therefore be less for all points of time prior to the end of the planning period as a result of the centralised decision in relation to the size of consumption in a decentralised economy without any state interference.

The centralised decision results in an optimum solution in which production, and consequently the amount of discharge, increases over time up to the production and discharge level in the perfect competitive economy. The optimum solution, in other words, results in growth, whereas free competition results in constant production or zero growth. It follows that the accumulated waste is less at any point of time in the optimum solution compared with the market solution except at the initial point of time. Figures 8.1 and 8.2 illustrate this conclusion.

Figure 8.2 illustrates the fact that in the perfect competition solution the accumulation of waste increases linearly, whereas in the optimum solution it increases exponentially. Naturally, the optimal accumulation of waste will be less than the accumulation in the perfect competitive solution. In other words, this applies even though we have presupposed the somewhat ungenerous inheritance decision to the effect that $Z(T)$ can be anything we like.

Figure 8.2 presupposes that up to the initial year the economy has pursued a perfect competition development.

Figure 8.1: Optimal production developments x_t^{opt}, and the perfect competitive solution, x^{ma}

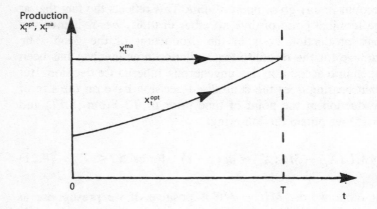

Figure 8.2: Optimal time development of accumulated waste and the corresponding development in a perfect competition society

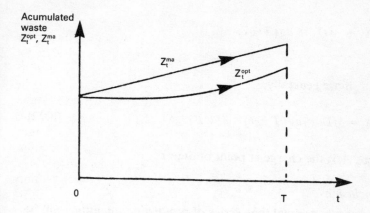

8.3.6 Implementation of the optimum solution

Let us assume that the collective body in the first 'partial period' of the planning period introduced an effluent charge A_0. Producers will then make an adjustment so that:

(i) Consumer price — effluent charge = MC and consumers will adjust themselves so that

(ii) consumer price = MC.

From (i) and (ii) it follows that

$$MU_0 = MC_0 + A_0 \qquad (8.22)$$

where MU_0 and MC_0 are marginal utility and marginal cost in the first period of the total planning period.p181

From (8.17) we see that if

$$A_0 = MD_0 = agT,$$

where T is the number of remaining periods, then the condition (8.22) will agree with (8.17). Producers will be guided to produce the production quantity that is socially optimal. If

181

this agreement is to continue throughout the period, it will be seen from (8.17) that the charge for the next 'period' will have to be set at

$$A_1 = MD_1 = ag(T-1)$$

etc.

Or, more generally,

$$A_t = MD_t = ag(T-t) \tag{8.23}$$

where A_t is the charge at point of time t.

Conclusion

The socially optimal time series of production quantities will be realised if a charge per unit produced is introduced at the point of time t equal to $ag(T-t)$. In other words, the charge will be variable. It is greatest to start with in the planning period, and decreases towards zero.

The charge can therefore not be calculated merely by taking into account the additional damage a hypothetical change in the accumulation of residuals generates at a given point of time. A charge must reflect future damage.

The time development for the optimal charge is shown in Figure 8.3.

Figure 8.3: Optimal charge development

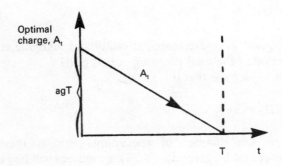

8.3.7 Appendix to section 8.4

I. The optimum problem without discounting the utility flows and without a biological breaking down of discharge substances

The optimisation problem discussed in 8.3.5 was

$$\max_{x(t)} \int_0^T \left(U(x(t)) - c(x(t)) - G(Z(t)) \right) dt \qquad (P)$$

Given

$$\dot{Z}(t) = a \cdot x(t) \qquad (1)$$
$$x(t) \geq 0 \qquad (2)$$
$$Z(0) = Z_0 \qquad (3)$$
$$Z(T) \text{ free} \qquad (4)$$

The optimisation problem P is one that can be solved with the aid of the theory for optimal control.

Sentence 1. Let $x = x^*(t)$ be the choice of time function for the control variable on $t \varepsilon [0, T]$ which solves the maximisation problem P. An auxiliary variable $p(t)$ then exists:

(i) so that for each t, $x^*(t)$ maximises:
$H(Z(t), x(t), p(t)) = U(x(t)) - c(x(t)) - G(Z(t)) + p(t) \cdot ax(t)$; for given p and Z and where $p(t)$ satisfies

(ii) $\dot{p}(t) = -\dfrac{\partial H}{\partial Z}$

evaluated at $x = x^*(t)$ and $Z = Z(t)$
(iii) transversality condition $p(T) = 0$.

The H function is called the Hamilton function. We can see that this is equal to the current economic surplus $\pi(x, Z) = U(x) - c(x) - G(Z)$, plus the increase in value of the accumulated waste $= p \cdot \dot{Z}$. The variables refer to the point of time t. $p(t)$ can be interpreted as a price at a point of time t associated with the accumulated waste.

The maximisation rule in Sentence 1 gives as a result the necessary condition

$$\frac{\partial H}{\partial x} = \frac{\partial \pi}{\partial x} + p \cdot a = 0 \qquad (5)$$

(given that $x = x^*(t) = 0$ is not optimal). (5) can also be transformed to

$$U'(x^*(t)) = c'(x^*(t)) - ap(t) \text{ for all } t. \qquad (6)$$

(6) expresses the fact that the control x at every point of time must be adjusted so that the immediate marginal gain U' is exactly balanced by the immediate marginal cost c' and the value calculated at the point of time t of the addition in accumulated waste, a, lasting as far as T, of increasing x marginally. a, of course, is equal to $d\dot{Z}/dx$.

$-ap(t)$ can be interpreted as an extra, external, marginal cost resulting from increasing production by one unit. The righthand side of the equation in (6) can be interpreted as the marginal social costs of increasing production at t. The difference between the private marginal cost c' and the social is represented by the term $-ap(t)$. Intuitively, $p(t)$ should be negative. We shall soon discover that this is, in fact, the case.

Can $p(t)$ be given the price interpretation we have used? It can. As it is applied in the theory for optimal control it is defined by

$$p(t) = \frac{\partial J(Z(t))}{\partial Z(t)}$$

where

$$J(Z(t)) = \max_{x(r)} \int_{t}^{T} \pi(x(r), Z(r)) dr$$

$p(t)$, in other words, is somewhat imprecisely equal to the addition to the economic surplus at point of time t resulting in an increase 'by one unit' of the accumulation of residuals at t. $p(t)$ therefore expresses the marginal contribution of the variable Z at the point of time t of the long-term economic surplus accumulating as from t. $p(t)$ therefore is very much in the nature of a shadow price. (The accumulation Z at point of time *t* along the path can be interpreted as a constraint originating in the past.)

Equation (ii) in Sentence 1 can be interpreted as follows: take as your starting point the Hamilton function; add 'capital gains' or, as Z is somewhat unacceptable, capital loss $\dot{p}Z$. Call the new H-function $H^*(x,Z)$.

$$H^*(x,Z) = \pi(x,Z) + p\dot{Z} + \dot{p}Z.$$

By maximising $H^*(x,Z)$ with respect to x and Z for given values of p, we see that the necessary condition (5) occurs, but also equation (ii) in Sentence 1.

$$\frac{\partial H^*}{\partial Z} = \frac{\partial H}{\partial Z} + \dot{p} = \frac{\partial \pi}{\partial Z} + \dot{p},$$

since Z is independent of Z. $\partial H^*/\partial Z = 0$ gives us (ii) precisely.

In other words, the maximum principle tells us that if possible the decision unit at an arbitrary point in time must adapt both the control and the state variable, so that the current economic surplus, plus the increase in the value of the accumulated waste, is maximised. 'If it was possible' means that it *is* impossible at a given point of time to change the size of a stock, here Z. The reasoning has therefore been included in order to explain the maximum principle.

So far we have only considered necessary conditions. The conditions mentioned will also be sufficient if

$$\max_{x} H(x,Z,p) = H^0(Z,p)$$

is a concave function of Z for given values of p and t. What is also demanded is that $H(x,Z,p) = U(x) - c(x) - G(Z) + p \cdot ax$ is a concave function of Z, given p,t and (5). We can easily discover that $\partial H^0(Z,p)/\partial Z = -G'_z = -g < 0$. According to the assumption made about the functions involved $\partial^2 H^0/\partial Z^2 = 0$.

$H^0(Z,p)$, in other words, is a concave function of Z. If $G(Z)$ had been a strictly convex function of Z, in other words $G' > 0$, $G'' > 0$, then $H^0(Z,p)$ would be a strictly concave function of Z.

185

Let us now collect the results we have arrived at:

$$\dot{Z}(t) = ax^*(t) \qquad (1)$$
$$x^*(t) > 0 \qquad (2)$$
$$Z(0) = Z_0 \qquad (3)$$
$$U'(x^*(t)) = c'(x^*(t)) - ap(t) \qquad (6)$$
$$\dot{p}(t) = + g \text{ (according to (ii) in Sentence 1)} \qquad (7)$$
$$p(T) = 0 \text{ (according to (iii) in Sentence 1)} \qquad (8)$$

From (1), (3), (6), (7) and (8) we can determine the time development for the three unknowns x^*, Z and p. From (7) and (8) we get

$$p(t) = p(T) - \int_t^T g \, dt = -g \cdot (T-t) \qquad (9)$$

Since $g > 0$ we have $p(t) < 0$ for all $t < T$.

From (6) and (9) we get

$$U'(x^*(t)) = c'(x^*(t)) + ag(T-t) \qquad (10)$$

g is equal to what above has been called marginal damage due to pollution. a is equal to the increase in accumulated residuals resulting from an increase of production by one unit. This increase in accumulated waste will, given our assumptions, last from the point of time of production and from then on. The marginal damage resulting from increasing production by one unit at point of time t will then be:

$$MD(t) = ag(T-t)$$

As mentioned in section 8.3.6, an economy based on decentralised decisions, but without corrective market guidance, might be described by

$$U'(x(t)) = c'(x(t)) \qquad (11)$$

From (10) and (11) it follows that

(i) $x(t)$ determined in (11) is constant over time.

(ii) $x(t)$ is greater than $x^*(t)$ for all values of t apart from $t = T$.

(iii) For $t = T$ we get $x = x^*$. This is due to the terminal condition: $Z(T)$ free, and due to the transversality condition $p(T) = 0$ derived from this.

As shown in section 8.3.7, the market mechanism can be governed so as to generate optimal production quantity (and therefore optimal pollution level) by introducing an effluent charge per produced unit of the good defined by

$$A(t) = a\,g(T-t) = -\,a\,p(t) \tag{12}$$

Producers will then maximise profit: $Q(t)x(t) - c(x(t)) - A(t)x(t)$.

$Q(t)$ is the product price for every t. When consumers at the same time maximise utility under a given budget condition, condition (10) will be satisfied.

In other words, (12) implies a charge depending on $T - t$, namely, the remaining period of the planning period T. The charge is greatest to begin with, decreasing towards zero when $t = T$. This is due to the transversatility condition.

If we differentiate through (6) with respect to t, using (7), we get:

$$\dot{x}^*(t) = \frac{ag}{c''\,(x^*(t)) - U''\,(x^*(t))} \tag{13}$$

This increment is positive for all $0 \le t \le T$ according to assumptions made earlier.

If we call the constant production quantity generated by the uncorrected decentralised market economy $\bar{x}$ ($\bar{x}$ is determined by (11) for all values of t), we can see from (10) and (11) that

$$x^*(t) - \bar{x} < 0 \quad \text{for all } 0 \le t < T \tag{14}$$

and that

$$x^*(T) - \bar{x} = 0 \tag{15}$$

If we call the associated accumulation of residuals respectively $Z^*(t)$ and $\bar{Z}(t)$, we shall have no difficulty in discovering that

$$Z^*(t) - \bar{Z}(t) < 0 \quad \text{for all } 0 < t \le T \tag{16}$$

187

and

$$Z^*(0) - \bar{Z}(0) = 0 \tag{17}$$

II. An extended optimisation problem with discounting and biological breaking down

There are several extensions of this problem that might claim our attention:

(1) Introduce the question of whether the good should be produced.
(2) Introduce several sectors of the economy and several types of residuals.
(3) Introduce the possibility that in the production of the good considered, a choice may exist between production processes and the amounts of residual discharged per produced unit, and the recipient.
(4) Introduce natural depreciation of accumulated residuals.
(5) Introduce other qualities of preference structure, for example, the fact that the utility flows at different points of time are accorded different weight.

We shall now briefly consider the last two points. We shall assume that the integral to be maximised is now

$$\int_0^T e^{-rt} [U(x(t)) - c(x(t)) - G(Z(t))]dt$$

where r is a social discount rate. We assume that there is a single discount rate, $r > 0$.

Eq. (1) above is replaced by

$$\dot{Z} = ax - \delta Z \tag{1a}$$

where δ is a depreciation factor. (A quantity Δ of residual disappears exponentially at rate δ owing to Nature's own capacity for breaking down.) These are the only amendments to problem P. The Hamilton function will now be

$$H(x,Z,p,t) = e^{-rt} [\pi(x,Z)] + p(ax - \delta Z) =$$
$$e^{-rt} [\pi(x,Z) + q(ax - \delta Z)]$$

188

where π is the previously defined current social surplus and $q = pe^{rt}$. This definition of q means that the content of the last bracket may be compared with current magnitudes.

Instead of (5) above we now get

$$\frac{\partial \pi}{\partial x} + q \cdot a = 0 \tag{5a}$$

and instead of (ii) in Sentence 1 we get

$$\dot{q} + \frac{\partial \pi}{\partial Z} - \delta q = rq \tag{iia}$$

We shall soon discover that $q(t) < 0$. The left-hand side may then be said to express capital gain $\dot{q}(t)$ resulting from a reduction at point of time t of the accumulated waste by one unit, plus the net marginal profit (gross profit adjusted for the depreciation value of the environmental stock per unit of the stock) resulting from reducing the accumulated residuals slightly, or not very exactly by one unit at point of time t. (iia) tells us that this total marginal gain must be equal to the interest rate r on the value of the accumulated residuals calculated positively (that is, on $-qZ$). The sufficient conditions for maximum are still fulfilled in so far as $-G(Z)$ is a concave function of Z.

As $p(T) = 0$ implies $q(T) = 0$, we derive from this and (iia):

$$q(t) = \frac{-g}{r + \delta} [1 - e^{-(r+\delta)(T-t)}] \tag{9a}$$

which is less than zero for all $0 \le t \le T$.

The parallel to the charge in (12) will then be:

$$A(t) = \frac{ag}{r + \delta} [1 - e^{-(r+\delta)(T-t)}] \tag{12a}$$

(12a) can be written

$$A(t) = ag \int_{0}^{T-t} e^{-(r+\delta)\tau} d\tau \tag{12b}$$

The correction to be introduced at every point of time in the decentralised economy will then be a charge per produced unit equal to the increase a in the accumulated waste resulting from an increase in production by one unit at point of time t, (a), multiplied by marginal damage (g) inflicted by the accumulated waste. This product will be discounted by a factor $r + \delta$ and where the period of discounting is equal to the remaining part of the planning period.

From (12a) follows:

$$\frac{\partial A(t)}{\partial n} = \frac{-ag}{n}[1 - e^{-n(T-t)} \cdot (1 + n(T-t))] \tag{18}$$

where $n = r + \delta =$ 'gross rate of interest' = rate of interest + rate of depreciation.

The contents of the brackets will be positive if

$$1 > e^{-n(T-t)}(1 + n(T-t))$$

That is, if

$$\frac{1}{\theta}\frac{1}{n}(e^{n\theta} - 1) > 1 \tag{19}$$

where $\theta = T - t$.

Eq. (19) is fulfilled for all $\theta > 0$ (namely, for $0 \le t \le t$) as (19) is equivalent to

$$\frac{1}{\theta}\int_0^\theta e^{n\tau}d\tau > 1 \tag{20}$$

and where the left-hand side of (20) expresses the mean value or average ordinate for the function $e^{n\tau}$ between $\tau = 0$ and $\tau = \theta$. This mean value must lie between the smallest and greatest values the function $e^{n\tau}$ obtains between 0 and θ. In so far as the smallest value is 1 (for $\tau = 0$), (20) and consequently (19) are fulfilled for $\theta = 0$.

From (18) it follows that $\partial A(t)/\partial n < 0$ for all $t \, \varepsilon [0,T]$. In other words, the charge is less for every point of time, the less weight one attaches to the future (the greater r is) and/or the better the breaking down is in Nature (the greater δ is).

(14)–(16) also apply when discounting and depreciation are included. From the above, however, it follows that the negative difference

$$x^*(t) - \bar{x}(t) \text{ for all } 0 < t < T$$

is less numerically, the higher $r + \delta$ is, and consequently that the negative difference

$$Z^*(t) - \bar{Z}(t) \text{ for all } 0 < t \leq T$$

is numerically less, the higher $r + \delta$ is.

REFERENCES AND FURTHER READING

d'Arge, R.C. (1971): 'Essay on economic growth and environmental quality', *Swedish Journal of Economics*, 73, 25-41

d'Arge, R.C. and R.G. Kogiku (1973): 'Economic growth and the environment', *Review of Economic Studies*, 40, 61-78

Forster, B.A. (1977): 'Pollution control in a two-sector dynamic general equilibrium model', *Journal of Environmental Economics and Management*, 4, 305-12

Førsund, F.R. and S. Strøm (1974): 'Industrial structure, growth and residual flows', in J. Rothenberg and J.G. Heggie (eds): *The management of water quality and the environment*, Macmillan, London, 21-69

Haavelmo, T. (1970): 'Some observations on welfare and economic growth' in *Induction, growth and trade*, essays in honour of Sir Roy Harrod, Clarendon Press, Oxford

Keeler, E., M. Spence and R. Zeckhauser (1972): 'The optimal control of pollution', *Journal of Economic Theory*, 4, 19-34

Strøm, S. (1972): 'Dynamics of pollution and waste treatment activities', *Memorandum*, Department of Economics, University of Oslo, May 10

9

Environmental Costs, Irreversibility and Uncertainty

9.1 INTRODUCTION

In this chapter we will concentrate on the possible environmental damage that might occur when certain areas are developed for industrial purposes. A well known example is the regulation of water courses located in recreational areas. Environmental costs related to such regulation of areas reveal several characteristics:

(1) Costs are a result of the alternative use that could be made of these areas either being reduced in scope and/ or in quality.

(2) Common areas are involved. The goods and service supplies which are reduced either in scope or in quality in this way acquire the nature of common goods. In most cases no prices exist for these goods, and in any case it is difficult to calculate the willingness to pay among potential users of the area. Some will also be willing to pay today for the *option* of being able to enjoy recreational services in the future. Irrespective of whether they actually use these services in the future or not, an economy aiming at social efficiency ought to take these option values into account when making plans for the economy.

(3) In some cases irreversible effects may be involved. Discharge of residuals may result in tolerance thresholds being exceeded, in such a way that the natural environment is exposed to lasting change. In the case of hydropower development, damming and the regulation of

watercourses may change the character of the country-side, *inter alia*, as a result of erosion. An ecologist will probably be more likely than an economist to maintain that irreversible damage has been done, since the latter will be more concerned with the more general facilities for use by human beings offered by the area, and in addition he will be more aware of the fact that, by means of investment at any rate, important features of the natural setting may be restored.

(4) Uncertainty is involved. In the first place this may involve uncertainty with regard to the physical effects of interference with the environment. In the second place there may be uncertainty with regard to the value of the goods that have been lost. A typical feature of the uncertainty associated with interference with the environment is that information on expected physical conditions and willingness to pay 'tomorrow' depend on realisation 'today'. After some time has passed and as decisions are taken, the decision-maker will receive more and better information.

In continuing our investigation we shall confine ourselves to the development of hydropower. Power development involves environmental disadvantages caused by the regulation of water-courses, the damming of lakes, the discharge of water, the building of transmission lines, etc. Environmental costs will occur if these interferences with the environment reduce the utilisation that others make of the area today or will be making in the future. Alternative utilisation is associated with other forms of economic exploitation or amenities. Under the heading of other economic forms of economic exploitation might be mentioned agriculture, forestry, inland fisheries, coastal fisheries, as well as industry and the service trades.

In making decisions on power development certain environmental disadvantages are taken into account.

According to the practice pursued in Norway today, the power developer is obliged to restock fish, build salmon ladders, undertake afforestation, develop tourist paths, etc.

Apart from the measures the developer himself is to set in motion, he must in many cases pay compensation for various forms of private economic loss. Compensation of this kind may involve covering increased irrigation costs, sewage expenses, and various other disadvantages that the regulation of a water-course may impose on the property owner.

In spite of this it is obvious that most watercourse regulation projects entail major or minor acts of interference with the natural environment for which the power developer is not forced to pay, and which for this reason are not included in development costs. These losses may be those that affect the general public as a result of a reduction in the scope or value of amenity areas. More direct economic loss, for which there is no legal address, may also be involved.

It must be pointed out that at times power development also involves advantages that do not accrue to the power developer. An example of this is that the regulation of a watercourse may in some cases at the same time provide a guarantee against flooding. Another example is that constructional roads may provide better access to amenity areas, or facilitate the utilis-ation of pasturage. This advantage, howeve, may be regarded by some people as yet another disadvantage of power develop-ment. We shall concentrate below on the recreational use that can be made of mountain, moor and watercourse. Our starting point will be that a conflict exists between the use of common areas for hydropower production and for recreational purposes.

9.2 THE DEMAND FOR RECREATION SERVICES: A PUBLIC GOOD EXAMPLE

The main problem involved in quantifying the costs involved in reduced recreational facilities is that the recreational goods with which we are here dealing are *public goods*. This means that a person's use of the good does not adversely affect other persons' use of that good. If we ignore the phenomenon of overcrowding, the recreational utilisation of mountain, moor, and watercourse will be a good that fits this definition.

What we are endeavouring to calculate is people's willingness to pay for a public good. In Figure 9.1, by way of illustration, we are assuming that the population consists of three persons. Each of them has maximum willingness to pay for the public good we are considering, in this case concretely equal to the country's total recreational areas that *may* be affected by power development (mountain, moor and watercourse). Maximum willingness to pay is what the individual is willing to pay when the public good involved is about to disappear for good and all. This maximum willingness to pay is not infinitely great, partly because people can manage to live without recreational goods, and partly because as a rule other recreational facilities exist (lowland, coast, other countries, etc.) to compensate for what is about to vanish. This emphasises the fact that willingness to pay for the common recreational area we are considering depends on the scope and availability of other recreational values. Each of the persons concerned, too, will possess a saturation amount of the public good under consideration, which means that none of these persons is prepared to pay anything for increasing the country's total recreational area in mountain and moor, if this exceeds a certain size.

It will be seen from Figure 9.1 that the community's total demand curve is the vertical sum of the individual demand curves. If the public good, that is, the country's total recreational area, possesses a scope equal to x^0, Person 1 will have a marginal willingness to pay equal to P_1, Person 2 one equal to P_2, etc. Willingness to pay for society as a whole equals the sum of these individual willingness to pay. This is an important factor

Figure 9.1: Demand for a public good or the community's marginal willingness to pay for preservation of the country's recreational area in mountain and moor

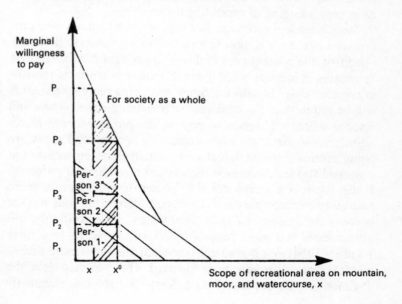

for a public good, and one that distinguishes it from a private good: marginal willingness to pay for a public good is equal to the sum of individual marginal willingnesses to pay.

When the total recreational area in Norway, say, amounts to x^0, marginal willingness to pay for the community as a whole is thus equal to P_0, which is to say:

$$P_o = P_1 + P_2 + P_3 \tag{9.1}$$

If the public good is less in area than x^0, the community's marginal willingness to pay will be greater than P_0, and conversely if the area is greater than x^0. It will be seen that when the scope of the recreational area as a whole in the country has been reduced to x, the community's marginal willingness to pay has risen to $\underline{P}$. This is important: the more that has already been taken of a country's recreational areas, the higher is the community's marginal willingness to pay in order to preserve what is left. This emphasises the importance of considering the country's total remaining recreational areas, when the value of a specific act of interference with the natural environment is to be assessed. If this is not done, the danger exists of being oriented too much on a bit-by-bit basis: this means that it is important at all times to keep up to date a survey which shows the scope of remaining recreational areas, in the same way as power developers so far have kept informed of remaining watercourses.

Since a tendency exists in Norway, admittedly not a very pronounced one, for the most favourable watercourses to be developed first, the consequence of this is that marginal costs in power development increase when the hydropower potential in remaining watercourses is reduced. Since remaining recreational areas will be reduced at the same rate as remaining watecourses, and since marginal willingness to pay for the preservation of recreational areas increases when remaining recreational areas are being reduced, the marginal costs that include environmental costs will also for this reason increase, as development proceeds. If this factor is not considered in power development, as more and more watercourses are developed, demonstrations against power development will take place which, *quite justifiably*, will attract more and more people! It will, of course, be seen from Figure 9.1 that as x is gradually reduced, more and more people will feel that values have been forfeited. They will then 'join' the P-function (see section (9.2) below). We shall subsequently note that there are other reasons, too, why willingness to pay increases over a period of time.

Generally speaking, we can therefore say that the community's marginal willingness to pay for the preservation of recreational areas increases when remaining recreational areas are reduced, that is:

$$P = P(x) \tag{9.2}$$

and where the P-function is such that P increases when x is reduced. We note from (9.1) that $P(x)$ consists of individual marginal willingnesses to pay. In our presentation we have simplified matters in such a way that each area, irrespective of topographic and geographic situation, is identical from the point of view of recreational amenities. This, of course, would not be the case in reality. This is a complication which can conveniently be dealt with in concrete analyses, but which we shall ignore here.

If power development not only involves a marginal reduction in the country's recreational areas, but also involves large tracts of country, the marginal willingness to pay will not provide sufficient information on what this interference with the environment costs. From Figure 9.1 we note that if the recreational areas are reduced from x^0 to $\underline{x}$, the community suffers a loss equal to the size of the shaded area in the figure. If we call the loss $V(\underline{x}, x^0)$, we get

$$V(\underline{x}, x^0) = \int_{\underline{x}}^{x^0} P(x)\,\mathrm{d}x \tag{9.3}$$

The connection between remaining recreational areas and hydropower potential might be measured in such a way that $P(x)$ is expressed in pence/kWh. $V(\underline{x}, x^0)/(x^0 - \underline{x})$, which is the average loss of utility resulting from the change from x^0 to $\underline{x}$, will in this way also be expressed in pence/kWh. For this reason $V(\underline{x}, x^0)$ is measured in pence.

The value of future use of the country's recreational areas depends on how the willingness to pay for this special recreational good rises in relation to 'all other prices'. These 'other prices' may be limited to the prices included in the other cost calculations associated with hydropower development and electricity prices, since the object of these estimates is to undertake an investment calculation, in which total development costs are compared with electricity prices. The demand for *private recreational goods* generally increases more rapidly than any demand in a growing economy. Generally this will result in prices of such products rising in relation to other prices. One of the reasons why this has not always taken place is that technical progress in the production of these products moderates the rise in price that would otherwise have taken place. Even though a country's recreational areas are not the same as any private recreational good, there are nevertheless certain common features. An important difference is that it is not possible to ensure any technical progress in the 'production' of the common recreational good. At the best of times it simply exists. New areas (natural or synthetic) can, of course, be discovered which will result in a negative shift in the maximal willingness to pay for preserving these areas.

A simplified market model will explain what happens. Let us assume a closed economy, where two consumer goods exist, a private good called an industrial good and a common recreational good called a natural good.

The natural good exists in a given and constant amount. The quantity of the natural good is conceived as a flow of services from the natural environment.

Work is the only input factor in the production of the industrial good. The input of real capital, energy, and raw materials will thus be ignored. These input factors may be considered to be 'lurking in the background'. The production function in the industrial sector is therefore assumed to involve diminishing returns to scale. The production functions are assumed to have positive, exogenously given, shifts over a period of time, so-called technical progress. We shall ignore the fact that it is necessary to make use of common areas in order to carry on industrial production.

Employment trends over time are assumed to be exogenously given, and are presupposed to follow population trends. Wage rates will then play a role in ensuring a balance between this exogenously given employment and the labour force required by the profit-maximising industrial sector. The surplus in this activity, plus gross income of the sale of the natural good (which is equal to net income) plus wage income in industrial activity, accrue to the workers. These maximise the utility which depends on the amount of the industrial good and the amount of the natural good. All income is spent in purchasing these two goods. No queues in the consumptions of the natural good are envisaged. In other words, indirect effects are ignored. Let X_F be the quantity of the natural good and X_1 the quantity of the industrial good, and p_F and p_1 the corresponding prices; r is the available income of the individual consumer; N is both an expression of the size of the population and of the labour force; w is the wage rate. Our model is then:

$$X_F = F(\frac{P_F}{P_1}, \frac{r}{P_1}) \cdot N \tag{i}$$

$$\frac{r}{P_1} = \frac{X_1}{N} + \frac{P_F}{P_1} \frac{X_F}{N} \tag{ii}$$

$$X_1 = e^{bt}(f(N)) \tag{iii}$$

$$\frac{\partial X_1}{\partial N} = \frac{w}{P_1} \tag{iv}$$

$$N(t) = N(0)e^{nt} \tag{v}$$

$$X_F(t) = \bar{X}_F \tag{vi}$$

$F(\cdot)$ is the demand function for the individual. The arguments in this function, as a result of the utility maximisation assumption, are relative prices and real income. The price of the industrial good is used to deflate all nominal magnitudes. $F(\cdot)$ · N gives the total demand for the natural good in the economy. (ii) tells us that real income *per capita* is used for the purchase of these two goods. Note that there is no room to insert the earning of income equal to real income *per capita* as an independent equation. (iii) is a production function for the industrial good. $f' > 0, f'' < 0, b > 0$. (iv) is a necessary condition for profit maximum, and in this model indicates demand for labour written in an implicit form. (v) is the exogenously given employment trend. (vi) tells us that the supply of the

199

natural good is given and constant over time. Unknowns in our models are:

$$X_1, X_F, N, \frac{P_F}{P_1}, \frac{w}{P_1} \cdot \frac{r}{P_1}$$

According to the counting rule the model is determined. If we assume that a solution exists for the model, the model will in other words determine sales of the two goods, the relative price P_F/P_1, real wage w/P_1, and real income per capita, r/P_1.

The problem we have to solve is what will happen to the relative price.

$$\dot{P}_F/P_1 = p \text{ over time.}$$

If we differentiate through the model with respect to t we shall find:

$$\frac{\dot{p}}{p} = (ba + nL)\frac{E_F}{(-\varepsilon_F)} + \frac{n(1-E_F)}{(-\varepsilon_F)} \tag{vii}$$

$\dot{p}/p$ is the growth rate for the relative price p_F/p_1; b is the technical rate of progress in the production of the industrial good; a is the budget share of the industrial good in the economy of the representative individual; n is the growth rate of employment (and of the population); L is the wage income's share of national income, that is, $L = [(w \cdot N)/(r \cdot N)] \cdot E_F$ is the income elasticity of the natural good and ε_F is the direct Slutsky elasticity of this good.

ε_F is negative, owing to the assumption that the indifference lines curve towards the origin. The numerical value of ε_F is less, the smaller the substitution possibilities are between natural good and industrial good. A more restrictive, though nevertheless reasonable, assumption is that $E_F > 1$. Owing to the two-good specification it then follows that $E_1 < 1$. The natural good has a touch of luxury as compared with the industrial good. Partly as a result of this we shall assume that the budget share of the industrial good is big, but declining.

The first term on the right-hand side in (vii) represents the effects of economic growth. The factors that cause this growth are technical progress expressed by rate b and growth in employment expressed by rate n. Economic growth, in other words, will result in people's willingness to pay for the natural good to grow over time. The greater the economic growth, the more the price of the natural good rises in relation to the price of the industrial good. If the production of the industrial good takes place at the expense of the natural good, the economic growth in itself will result in a damping of growth. This presupposes that mechanisms are established in the economy which ensure that the indirect effects of industrial production are taken into consideration. More and more clamant demands for protection of the natural environment can be understood in the light of the foregoing reasoning. We can observe, too, that the first term in (vii) is greater, the greater the income elasticity E_F is and the smaller the substitution possibilities between industrial and natural good are. The first term in (vii) does not become constantly greater over a period of time; presumably it will become less over time. Growth rates for the relative price will also probably decrease over time. In part this is due to the circumstances already mentioned, which will dampen the growth of industrial production, and in part (in our model here) an element will be introduced of declining profits in industrial production, and in part it will be due to the fact that as people gradually become richer, the parameters on the demand side will be changed. In view of the fact that it is assumed that the income elasticity of the industrial good is less than 1, we shall be able to calculate that the budget share a will be less over time. In addition, we must take into account that $E_F/(-\varepsilon_F)$ most likely is reduced over time. As people become richer and richer, we must expect the substitution possibilities to become greater, that is, ε_F becomes greater in absolute value.

The second term on the right-hand side of (vii) is negative, since E_F is assumed to be greater than 1. The reason for this is that the increase in demand, resulting from the increase in population, considered in isolation, will in the first place be directed towards the more 'necessary' good. The increase in population will therefore result in the price increase for the natural good being reduced.

The necessary condition for the entire right-hand side of (vii) to be positive is that

$$E_F(ba - n + nL) + n > 0 \qquad \text{(viii)}$$

201

By inserting reasonable figures for b, a, n, and L, (viii) will be fulfilled. ba is of the order of magnitude 0.03, $n \cdot L$ of the order of magnitude 0.01. The inequality in (viii) will then be nothing more than a demand that the growth in natural income per capita should be positive. Note that incomes from 'sales' of the natural good then have been added to incomes from the production of the industrial good.

There are therefore good reasons for estimating that willingness to pay for the preservation of recreational areas will increase in relation to other prices included in investment calculations associated with power development. An acceptable simplification is that the marginal willingness to pay increases by the same positive factor over time, irrespective of the scope or extent of remaining recreational areas. (All individual curves in Figure 9.1 change right across the diagram at the same speed over time.) If we call this factor g, the marginal willingness to pay at a future point of time t (that is, in t years' time) will be

$$P_t = P(x_t)(1+g)^t \tag{9.4}$$

x_t is the remaining recreational area at point of time t. Eq. (9.4) means that people's willingness to pay for protection of the natural environment increases by $100\,g$ per cent per annum in relation to other prices. On account of the simplification mentioned above as to how the curves in Figure 9.1 shift over time, we get

$$V_t = V(\underline{x}, x^0)(1+g)^t \tag{9.5}$$

In order to quantify people's willingness to pay for their common recreational good, there are two factors that must be quantified. In the first place we have the $P(x)$ function. As mentioned in connection with Figure 9.1, this is the sum of the 'willingness to pay functions' of the entire population of the country. When the $P(x)$ function is known, we also know the $V(\underline{x}, x^0)$ function. In the second place we need to know how the $P(x)$ function changes over time, that is, the growth rate g.

Direct method. In this method people are asked what they are willing to pay in order to have the use of a slightly bigger common recreational area or what they are willing to pay not to lose a little of their existing common recreational area. This provides an insight into the marginal willingness to pay for every single individual at a point equal to the existing common recreational area of the day. This provides knowledge of a point on the $P(x)$ function. Since we have assumed linear curves in Figure 9.1, it is sufficient to ask every person only two questions, in order to establish the curve. If we stick to this assumption we can, for example, ask for answers to the following two questions:

(1) Today the country's common recreational area is as follows (extent and other characteristics). How large do you consider that the country's recreational area should be? Do not take into consideration the fact that the areas can be used for purposes other than recreation.

Since there is no mention in this question of the cost of using the area, one can expect to get answers to the saturation amounts for the common recreational good, that is, the intersection with the horizontal axis in Figure 9.1.

(2) Suppose that it were no longer possible to utilise the country's mountains and moors for recreational purposes. Where would you then instead travel in the course of a year?

By discovering what the individual's annual extra expense would be, it would be possible to calculate what people are at least willing to pay for the common recreational good when it disappears completely, in other words the intersection on the vertical axis in Figure 9.1. By drawing lines for every person and adding up, we can discover what the most conservative $P(x)$ function is.

The weakness in this set-up is the hypothetical factor in the situation, whether one asks a question with regard to willingness to pay on the basis of the present-day scope of the recreational good or on the basis of other levels, as in the last two questions above. The hypothetical element lies in the fact that people do not pay and/or are faced with other binding restrictions. The answers are liable to be unrealistic, in the sence that the willingness to pay is overestimated.

If actual payments are introduced, the 'players' are motivated to pay less than they are actually wish. Each individual will bank on others paying ('What difference does it make if I fail to pay, or merely pay a small amount, so long as the others pay?'). Since everyone has a good reason for behaving in this way, the community's total willingness to pay for the recreational good will be underestimated.

Indirect method. This method involves discovering people's willingness to pay for common goods by means of observations of their purchases of the sort of private goods that are necessary in order to utilise the common recreational area. (Travel expenses to and from mountain areas, extra expenses involved in living and eating outside the home, expenses involved in purchasing equipment, such as rucksacks, boots, windproof clothing, etc., calculated as an annual expense, taking into account lifetime, rate of interest, and time spent in the mountains.) In this way we can reveal people's marginal willingness to pay for preserving the public good.

Converting to pence per kWh can be done by adding together all or parts of the expenses that people have incurred in the course of the year for visits to areas which can potentially be harnessed for the production of hydropower. According to holiday and open-air surveys carried out by the Central Bureau of Statistics of Norway in 1970, 4 million person-days were spent in 1969/70 on extensive tours on foot in the mountains.

Some of these person-days were quite clearly spent in other areas than those capable of providing hydropower, and it must also be assumed that some people also went for walks in regulated (developed) areas. On the other hand, other forms of touring took place in the mountains (skiing trips, short walks), even though these other tourists were clearly not to the same extent embarrassed by watercourse regulation measures.

A great many objections may be made to this approach: some have already been mentioned, but in addition we should like to draw attention to the following:

The use of the immediate environment does not involve costs, so the above method must be adapted by asking people in the immediate environment what they are willing to pay for retaining the environment unchanged. This will clearly prove more speculative, since what people actually pay is more tangible than what they say they are prepared to pay.

What people pay in the form of direct expenses in order to be able to walk in the mountains may be far less than what they are willing to pay for environmental protection. Moreover, as already noted in the introduction of this chapter there obviously are some that are willing to pay for the *option* of enjoying recreational facilities in the future.

These three factors tend to make us underestimate the willingness to pay when using this method.

Since we treat all areas alike, we presuppose that one area is a perfect substitute for another. In other words, it is a matter of no consequence to people where they can walk, provided they can walk in mountain country. This is clearly wrong, since the same willingness to pay would not apply to the preservation of the 'Jotunheim' range as opposed to the 'Svartisen' glacier. This can be dealt with by considering willingness to pay separately for each area.

As already mentioned, harnessing watercourses entails a *reduction* and not a *loss* of a recreational area. People may even have the sort of preferences that ensure that harnessing watercourses is not regarded as any disadvantage, for example some engineers may enjoy the sight of dam construction and slag heaps.

These last two factors involve overestimating willingness to pay for conservation of the national environment.

In order to quantify the growth rate g we can in principle proceed in the same way as when we establish the $P(x)$ function. It is necessary to obtain a time-series material which makes it possible to quantify shifts in the willingness-to-pay function. We do not possess material of this kind. On the basis of what we have already said about a growth economy, it is obvious that $g > 0$. A possible estimate can be obtained by analysis of the rise in real prices for private, marketed recreation goods. We can also refer to average investigations. In the above-mentioned investigation into holiday and open-air life it will be seen that enjoying open-air life increases in proportion to income.

9.3 IRREVERSIBLE INTERFERENCE WITH THE ENVIRONMENT UNDER FULL CERTAINTY

We shall use the term x_0 (measured in TWh) to indicate the present-day energy or initial capacity in the hydropower system. $\hat{x}$ indicates the energy level which makes it profitable to use thermal power stations in the further expansion of energy supply. This means that $\hat{x}$ is actually dependent on costs in the cheapest alternative to hydropower. Consequently $\hat{x} - x_0$ indicates the remaining economic exploitable — but not yet developed — hydropower capacity. As above, the volume of remaining recreational areas will be measured in terms of TWh. Consequently, remaining recreational areas in the conflict zone will be equal to $\hat{x} - x_0$. The development for hydropower purposes of remaining capacity will thereby increase hydropower production from x_0 to $\hat{x}$ TWh, but at the same time reduce recreational areas in the conflict zone from $\hat{x} - x_0$ to nought, $\hat{x} - x_0$ corresponds to x^0 in Figure 9.1.

Let $\bar{x}$ indicate the remaining water power capacity, so that

$$\bar{x} = \hat{x} - x_0$$

The amount of water that will be developed for energy purposes is denoted x_H and the amount allocated for recreational purposes is denoted x_R. This gives us

$$\bar{x} = x_H + x_R \tag{9.6}$$

Since precipitation and affluxion are stochastic magnitudes, all x's referring to hydropower are stochastic magnitudes. In the sequel we shall ignore precisely this point, since this is a complication that has no bearing on the principal conclusions we aim to arrive at.

The problem confronting a central planner will now be the following: the demand for electricity now and in the future is either given exogenously as a production programme or as an estimated time function depending on, *inter alia*, price and income development. The electricity price involved is in this case an equilibrium price, which means that production development and price must be decided on simultaneously. An alternative procedure is to allow the electricity price to be exogenously given. This will be the case if Norwegian power can

206

be sold on the North European market, while at the same time Norwegian deliveries constitute a small fraction of the total consumption in this market. Once again, in order to make our point as simply as possible, we can simplify matters by stating that the problem involves covering an exogenously given consumption of power in the future. The income from hydropower development will then merely comprise the costs saved by using hydropower instead of thermal power. By investing in power supplies hydropower production will in this way gradually be increased from x_0 to $\hat{x}$ TWh.

However, increasing power production from x_0 to $\hat{x}$ TWh will at the same time reduce the recreational areas on the conflict zone from $\bar{x}$ to nought. Thus, power development involves a loss of recreational income: this loss is one of the costs of hydropower development. It is therefore the task of the central planner to choose an investment programme in hydropower development which will ensure weighting the relative merits of the use of the areas involved between recreational purposes and hydropower production, namely, an optimal time development for x_H and x_R.

In solving this problem, the planner in accordance with our assumptions must take into account that power development involves irreversible interference with the natural environment. In the case of most economic decisions uncertain information with regard to the point of time for decision-making is involved. This means that in some cases one may, in time to come, regret what one has done. When the conditions involved in the decision prove reversible, it is possible to take such regrets into consideration by undertaking a fresh decision. Where irreversibility is involved, it is too late for regrets. If decisions involve irreversible conditions, ignoring this in making the decision will result in regretting more frequently than is optimal.

This statement also implies that it is, of course, possible to make a decision, while taking full account of the irreversibility that one will later regret. In a later section we shall discuss the consequences that uncertain information in fact entails, *inter alia*, for investment calculations, when decisions have to be made on the basis of irreversible conditions. We shall here confine ourselves to a discussion of decisions made under full certainty.

The central planner has been given two tasks: he can either gradually develop the areas for hydropower production or lease them for recreational amenities. It will be assumed that a development programme will be chosen such that discounted incomes from these two activities are maximised; given the conflict condition (9.6), given the irreversibility assumption that investments in power supplies are non-negative, and given the condition for a given development of power consumption that has to be covered.

This is a standard problem that can be solved within the theory of optimal control. We shall select a simpler, but for that reason also less precise, approach. Instead of dealing with the problem with the aid of a formal, dynamic analysis, we shall deal with the problem step by step starting with a static and reversible problem.

This means that at any point of time, t, we shall maximise the total value of utilising the areas for recreation and for hydro-power production. Thus at every point of time we shall determine an optimal magnitude x_H and x_R within the framework set by the remaining potential $\bar{x}$. We then come to the salient point: *If this optimal solution, constantly repeated over time, indicates that it is optimal to reduce hydropower production,* this means that initial conservation of certain parts of these areas would be optimal. Just how large the areas to be allocated for recreational purposes are can be decided with the aid of a present-value calculation. Since in the 'real world' uncertain information is involved, it is important that these calculations should constantly be revised.

When power development reduces recreational areas from $\bar{x}$ to x_R, this corresponds to the following loss of recreational income at a point of time t according to (9.5).

$$V_t = V(x_R, \bar{x})(1+g)^t$$

or in continuous time

$$V_t = V(x_R, t) = V^*(x_R) e^{gt} \tag{9.7}$$

where the initial recreation area $\bar{x}$ is suppressed in the function for expository reason. According to (9.3)

$$\frac{dV^*(x_R)}{dx_R} = P(x_R) > 0 \tag{9.8}$$

$$\frac{\mathrm{d}^2 V^*(x_R)}{(\mathrm{d}x_R)^2} = P'(x_R) < 0 \tag{9.9}$$

The income from hydropower development can be arrived at in the following manner. Let c_v be the marginal cost equal to the average cost of the thermal power alternative to hydropower. In hydropower production we cannot assume that the marginal costs are constant. If we call this (long run) marginal cost c_H, it would be reasonable to suppose that it is an increasing function of the production quantity. This assumption is satisfied if the watercourses are of varying quality and if watercourses are developed in accordance with increasing costs. According to this, the surplus income, π_t, resulting from an increase in power supplies from level x_0 by means of hydropower development, instead of developing thermal power, will be

$$\pi_t = \pi(x_H, t) = c_V e^{vt} x_H - \int_{x_0}^{x_0 + x_H} c_H(x)\,\mathrm{d}x \tag{9.10}$$

The term e^{vt} will take care of changes in the unit cost of thermal power production over time. If, for example, the price of coal or natural gas rises in relation to the other prices, then $v > 0$. x_0 is the initial production capacity, which is given. We have assumed that the marginal cost function, c_H, will not change over time. A negative shift could have been caused by technical progress in power production or as a result of falling real prices for constructional activity caused by technical progress in the construction sector. Positive shifts could be caused by increasing real wages in construction work, etc.

$$\frac{\mathrm{d}\pi}{\mathrm{d}x_H} = c_V e^{vt} - c_H(x_0 + x_H) \gtreqless 0 \tag{9.11}$$

$$\frac{\mathrm{d}^2\pi}{\mathrm{d}x_H^2} = -c_H'(X_0 + X_H) < 0 \tag{9.12}$$

On this basis the static, that is, constantly repeated, optimisation task can be defined as follows:

209

$$\underset{x_H, x_R}{\text{maximise}} \ \pi_t + V_t = \pi(x_H, t) + V(x_R, t) \tag{9.13}$$

given (9.6)

$$\bar{x} = x_H + x_R$$

If we concentrate on an interior solution, the necessary condition for solving (9.13) will be

$$\frac{d\pi}{dx_H} + \frac{dV}{dx_R} \cdot \frac{dx_R}{dx_H} = 0$$

or according to (9.7), (9.8) and (9.11):

$$c_v e^{vt} = c_H(x_0 + x_H) + P(x_R) e^{gt} \tag{9.14}$$

From (9.13) we can see that the recreational income V_t is an opportunity cost in hydropower production. The equilibrium condition (9.14) tells us that at each and every point of time we must set aside sufficient for hydropower production that the total, that is, environmentally corrected, marginal cost in hydropower production will be equal to the marginal costs of the cheapest thermal power alternative. Eq. (9.14) determines implicitly x_H and x_R as functions of time.

From (9.14), by differentiating with respect to time t, and after some rearrangement, we get:

$$\frac{dx_H(t)}{dt} = \frac{vc_v e^{vt} - gP(x_R) e^{gt}}{c_H'(x_0 + x_H) - e^{gt} P'(x_R)} \tag{9.15}$$

Hence,

$$\frac{dx_H(t)}{dt} < 0 \text{ for all } t \text{ if } v \cdot c_v e^{vt} < gP(\bar{x}) e^{gt} \text{ for all } t$$

where $\bar{x}$ is the initial recreational area. Using (9.14) this last

210

inequality can be rewritten as

$$\frac{vc_H(x_0)}{P(\bar{x})e^{gt}} < g - v \qquad (9.16)$$

x_0 is the initial area developed in the past to give hydropower. Since all the terms to the left in the last inequality are constant except for e^{gt} in the denominator, the left side of the inequality will gradually decline over time towards zero. As t gets sufficiently large, the left-hand side will for practical purposes be equal to zero. Thus we get

$$\frac{\mathrm{d}x_H(t)}{\mathrm{d}t} < 0 \text{ provided } g > v. \qquad (9.17)$$

Reduced hydroelectric production over time would, in other words, prove optimal, as long as willingness to pay for recreational amenities increases more strongly over time than the increase in, for example, the price of coal, gas or oil (depending on which thermal power alternative is the cheapest).

This might be the case as far as Norway is concerned, since Norway is a growing net exporter of such sources of energy as oil and gas. A rise in the real price of these products would in this way ensure Norway increased revenue, which would, in particular, be reflected in a positive shift in the direction of recreational goods. When v increases, g will increase still more. Since hydropower development is a 'construction-intensive' activity and for this reason too a labour-intensive activity, it would be realistic to assume a positive shift in the cost function of hydropower over time, owing to rising real wages over time. Real wages, too, will probably rise more steeply, the steeper the rise in the price of oil and gas. Since it is relative prices that decide the adjustment in (9.14) this consideration, however, has already been taken care of. A positive v and a positive g mean that the thermal power cost and the 'recreational price' are assumed to rise relative to the price elements rising in the c_H function.

If (9.17) is fulfilled, this means that decreasing hydropower production and consequently increasing utilisation of recreational facilities could be optimal over time. Since our starting-

211

point was that water power development involves irreversible interference with the natural environment, we are led, in the static case, to a decision which involves determining, today and for all time, how much of the remaining area should be allocated for hydroelectric production.

In order to find a solution to the actual, dynamic and irreversible problem, we can proceed by way of three stages.

Stage 1

When (9.17) is fulfilled, considerations of irreversibility mean that the optimal x_H^{opt} is constant over time, and must consequently be equal to the solution we find in (9.14) by setting $t = 0$.

If this solution x_H^{opt} is equal to zero or is negative (in the latter case we should have to have $c_H(x_0) + P(\bar{x}) > c_v$); this means that it is optimal to maintain hydropower production at the initial level x_0. We should then either in advance be optimally adjusted or else it would be too late to undo the 'mistakes' of previous years.

If the solution is $\bar{x} > x_H^{opt} > 0$, this means that it would be optimal to extend hydroelectric production form x_0 to the level $x_0 + x_H^{opt}$.

Stage 2

Since our starting-point was an externally given time development for power demand to be covered by hydro and/or thermal power production, we can calculate, taking as our starting-point $x_0 + x_H^{opt}$, how many years in the future we are to expand by building new hydropower stations. If we assume that the externally given demand increases by the quantity $\Delta x(t)$ per time-unit, the period of time extending to the point where it would be profitable to change over to thermal power stations would be given by the following formula:

$$\int_0^T \Delta x(t) \mathrm{d}t = x_H^{opt} \tag{9.18}$$

If we say that the period comprises T years in which production is to increase by Δx_t every year, we can calculate the present value of each of these T projects as the years go by. It will then

212

be seen that the present value of each of these projects will be positive, but decreasing. The last project, that is, in year T, will have a present value which is exactly equal to zero. Present value calculations of the first T projects would, in other words, lead to the same conclusion as the result of a once-for-all optimisation, namely, a supplementary quantity of water power totalling x_H^{opt} TWh.

Stage 3

Once we have reached this production level, a further increase in hydropower production would not be optimal. Conservation of the remaining area, which may produce $\bar{x} - x_H^{opt}$, will then be optimal. If we call this remaining quantity z, and the capital investment required to develop extra hydropower capacity of this kind K_0, the following condition will have to be fulfilled if conservation is to be optimal (ρ is the social rate of discount):

$$\int_0^\infty V^*(z)\, e^{gt} e^{-\rho t} dt \geq \int_0^\infty c_v e^{vt} \cdot z e^{-\rho t} dt - K_0 \tag{9.19}$$

where ρ is the real social rate of discount.

We have then assumed that the operating costs of hydro-electric production are equal to zero. In order to harmonise with the tacit assumptions at the basis of the cost and income functions in the static case, we shall allow the lifetime for recreational incomes to be equal to the lifetime of hydropower stations. In order to simplify, we shall make this lifetime infinite, which is a reasonable assumption as far as recreational incomes are concerned, but not quite as reasonable where hydropower stations are involved. It is furthermore presupposed that all investment, K_0, takes place during the first year. We note that $V^*(z)$ is what we are willing to pay for conservation of recreational areas 'today'.

It will be seen from (9.19) that if $v > g$, then (9.19) cannot be fulfilled. This agrees with the conclusion that followed from the static set-up above. Thus, as above, if (9.19) is to be fulfilled, then we must have $g > v$. If, in addition, $g > \rho$, we shall straightaway see that (9.19) will always be fulfilled. If we multiply through (9.19) by ρ we get

213

$$\rho \cdot \int_0^\infty V^*(z) e^{gt} e^{-\rho t}\, dt \geq \rho \cdot \int_0^\infty c_v \cdot$$

$$e^{vt} \cdot z e^{-\rho t} dt - \rho K_0 \tag{9.19'}$$

We have here arrived at an interesting interpretation of the condition that characterises the optimal conservation quantity. The lefthand side of (9.19′) is merely the interest yield per annum of the present value of future recreational revenue, that is, the interest yield per annum of recreational capital. The first term on the righthand side is the interest yield of present value resulting from utilising the area for hydropower production. We should remember that power from the area is at all times evaluated at the thermal power cost $c_v e^{vt}$, since our starting-point is a given demand for power which is to be covered either by hydro or thermal power production. The last term on the righthand side is the annual user cost of real capital required if hydropower production is to be increased by z TWh. Eq. (9.19′) thus tells us that conservation of z TWh would be optimal if the yield per annum from leasing out areas for recreational use is greater than or equal to the net yield per annum of alternatively using the area for production of hydroelectric power.

An alternative method of interpreting (9.19) is obtained by making the opportunity costs of conservation equal to C. In this way C is equal to the right-hand side in (9.19). When $g<$ (9.19) can be written as follows:

$$V^*(x) \geq (\rho - g) \cdot C \tag{9.19''}$$

It is therefore optimal to exempt from power development z TWh, provided one is willing today to pay at least a net interest yield, equal to the discount rate minus the rate of price increase for recreational goods, on the cost of this conservation. We can see that the closer the rate of increase in future willingness to pay is in relation to the discount rate, the less is required in paying today to ensure that conservation should be optimal.

We shall demonstrate the use of formula (9.19″) by inserting realistic (1981) figures for Norway in 1979. $\bar{x}$ in South Norway (south of Salten in Nordland) is at present-day thermal power costs equal to 110 TWh. In 1979 approximately 80 TWh was

214

developed. Meanwhile, a number of power stations are being built; we can therefore consider the condition for ensuring conservation of the last 20 TWh, that is, $z = 20$ TWh $= 20$ mldr kWh. A fairly reasonable estimate of the investment required to develop these last 20 TWh would be approximately NoK 2.10 per kWh (1979 prices). This means that K_0 is NoK 40 billion. By way of simplification we shall assume that $v = 0$, that is, that thermal power costs do not rise in relation to cost elements in water power development over time. We shall assume that the discount rate is 7 per cent per annum while c_v is 0.15 NoK/kWh which corresponds to the present-day costs of coal-fired power stations (1979). This means that C will be equal to NoK860 million.

(9.19″) then tells us that conservation of these 'last' 20 TWh is optimal if

$$V^*(z) \geq (0.07 - g)\ 860 \text{ million NoK} \qquad (9.19''')$$

In Table 9.1 below we have calculated what people as a whole should at least be willing to pay today in order for this conservation to be optimal, given the alternative assumption for an

Table 9.1

Increase in future willingness to pay for recreation $100g$ (% p.a.)	What people as a whole would have to pay today in order for conservation of 20 TWh to be optimal, $V(20)$, millions of NoK (1979 prices)	What has to be paid, NoK per inhabitant today
0	60.2	15.05
0.01	51.6	12.90
0.02	43.0	10.75
0.03	34.4	8.60
0.04	25.8	6.45
0.05	17.2	4.30
0.06	8.6	2.15
0.07 and over	0	0

increase in future willingness to pay for recreation. We have also included a column showing what each inhabitant in Norway today would have to pay in order to ensure a conservation of this kind.

This brings us to the following general conclusions:

(1) Take a power-development programme which has to be covered either by hydro or thermal power as given. If willingness to pay for conservation is expected to rise more steeply over time than the price of oil/gas/coal, conservation of watercourses that would otherwise be capable of being economically developed would be optimal.

(2) The scope of conservation is characterised by the fact that the yield per annum of the present value of the areas for recreational purposes is exactly equal to the yield on the present value of the areas for hydropower purposes.

9.4 IRREVERSIBLE ENCROACHMENT ON THE ENVIRONMENT AND UNCERTAINTY

If encroachments on the natural environment are irreversible, already at the decision stage account must be taken of the fact that power development will mean that values will be lost for all time. The available information both on the extent of the damage and on future values is uncertain. Information, for example, on willingness to pay for recreation will generally improve as time goes on. This will be the case whether a watercourse is developed or not. If a watercourse *is* developed, knowledge of the scope of the damage will be acquired; if it is not developed, gradually more knowledge will be available of the frequency of visits and so forth. Had decisions been reversible, it would at all times and in time have been possible to take this growing body of information into account. Since decisions are irreversible, in order to ensure that the decision is optimal it is necessary to take into consideration the possible increase of information over a period of time, starting as early as the decision stage.

We shall show below that if the strategic nature of the decision problem in our situation is ignored, *all too frequently* one will be led to undertake irreversible decisions. Another conclusion that

216

may be drawn is that uncertainty, combined with irreversibility, involves a cost in itself.

In relation to the set-up above we shall make the following simplifications:

We shall consider two periods only, and a power project that can be developed in either the first or second period, if at all. The project is a small one, which means that we are operating with marginal income and costs, which are assumed to be independent of total power production and the recreational areas. All net income magnitudes are stochastic.

The following notation is used:

Δx = total amount of power that can be developed, TWh
Π_i = net income per unit of power produced in period i
V_i = net income per unit of the area used for recreational purposes in period i
Y_i = $\Pi_i - V_i$

In Π_2 and V_2 there is a 'concealed' discount factor. Π_i and V_i are stochastic magnitudes.

We have to arrive at a decision rule for period 1. It is assumed that the decision-making body will choose an irreversible project in period 1, $\Delta x > 0$ if the expected income from this project over the two periods is greater than the expected income when no irreversible project is undertaken, that is, when $\Delta x = 0$. Since the planner decides on the basis of expected magnitudes, he is risk-neutral.

The expected income when $\Delta x > 0$ in the first period we shall call R_+ and we shall call it R_0 when $\Delta x = 0$ in the first period.

R_+ is obviously equal to the expected income in period 1 accruing from sales of amount of power Δx_1 at a price Π_1 and expected power production income in period 2:

$$R_+ = E[\Pi_1 \Delta x + \Pi_2 \Delta x] \tag{9.20}$$

Preservation in period 1 postpones the decision whether to develop to period 2:

$$R_0 = E[V_1 \Delta x + \max[\Pi_2, V_2] \Delta x] \tag{9.21}$$

217

If we call ΔR, the difference $R_+ - R_0$, then the decision rule for development in the first period will be

$$\Delta R > 0 \Rightarrow \Delta x > 0 \tag{9.22}$$

$$\Delta R \leq 0 \Rightarrow \Delta x = 0$$

where

$$\Delta R = R_+ - R_0 = E[\Pi_1 \Delta x + \Pi_2 \Delta x - V_1 \Delta x$$

$$- \max[\Pi_2, V_2]\,\Delta x] \tag{9.23}$$

$$= E[(\Pi_1 - V_1)\Delta x + \min[0, \Pi_2 - V_2]\,\Delta x]$$

$$= E[Y_1 \Delta x + \min[0, Y_2]\,\Delta x]$$

We can distinguish between two cases:

Case 1. $Y_2 > 0$: $\Delta R = E[Y_1 \Delta x]$

Case 2. $Y_2 \leq 0$: $\Delta R = E[Y_1 \Delta x + Y_2 \Delta x]$ $\qquad$ (9.24)

It will be seen that if we could be sure that $Y_2 > 0$, it would be correct today to undertake the irreversible project $\Delta x > 0$ if only the expected difference in income between power production and recreational purposes today were positive. If, on the other hand, we were sure that the future difference in income Y_2 was negative or zero, it would be necessary to calculate the expected, total discounted income over both periods in order to decide whether an irreversible project should be undertaken today. It should be remembered that the discount factor is concealed in Π_2 and V_2. Since the difference in income in period 2 is now zero or negative, we can see that the difference in income in period 2 will be a cost involved in the irreversible project.

On the lines we have defined uncertainty, Y_2 is conditional on net income, Y_1, realised in period 1. We can then see from (9.22) and (9.23) that even though the decision-maker is risk neutral, we cannot insert at the point of time of decision expected values for Y_1 and Y_2 and then make a decision. We shall call a mistaken decision of this kind ignorance of uncertainty, and the approach will result in a discounted value which we shall call ΔR_{ig}.

$$\Delta R_{ig} = E[Y_1]\,\Delta x + \min[0, E[Y_2]]\,\Delta x \qquad (9.25)$$

ΔR_{ig} is the standard method of dealing with uncertainty in cost-benefit risk-neutral analyses without considering irreversibility. All uncertain magnitudes in an investment calculation are replaced by unexpected values. A decision rule generally used is

$$\{\Delta R_{ig} > 0 \Rightarrow \Delta x > 0\}$$

Comparing the difference in net income of developing or preserving in the first period we have from (9.23) and (9.25):

$$\Delta R_{ig} - \Delta R = \min[0, E[Y_2]]\,\Delta x - E[\min[0, Y_2]\,\Delta x] \qquad (9.26)$$

If we assume that both negative and positive values for net income Y_2 may occur, $P_r(Y_2 < 0) > 0$ and $P_r(Y_2 > 0) > 0$, it follows that

$$E[\min[0, Y_2]] < \min[0, E[Y_2]] \qquad (9.27)$$

When forming the expectation on the left-hand side of (9.27) the distribution for Y_2 is truncated at zero; positive Y_2-values are replaced with zero:

$$E[\min[0, Y_2]] = P_r(Y_2 < 0) \cdot E[Y_2/Y_2 < 0]$$

Since we have

$$P_r(Y_2 < 0) \cdot E[Y_2/Y_2 < 0] < E[Y_2]$$

$$= P_r(Y_2 < 0) \cdot E[Y_2/Y_2 < 0] + P_r(Y_2 > 0) \cdot$$

$$E[Y_2/Y_2 > 0], \qquad (9.27)$$

Returning to (9.26) we therefore have under our reasonable assumptions $\Delta R < \Delta R_{ig}$.

The use of customary investment calculations thus results, according to (9.26) and (9.27) in irreversible decisions too often being taken. If, for example, the following applies:

$$\Delta R < 0 < \Delta R_{ig}, \tag{9.28}$$

the correct calculation would result in the irreversible project being postponed in the first period ($\Delta x = 0$), whereas the ordinary standard investment calculation would result in an irreversible project being undertaken.

The difference

$$k = \Delta R_{ig} - \Delta R \tag{9.29}$$

may be called the cost in the event of uncertainty associated with an irreversible action.

Per se, uncertainty involves costs, even when the decision-maker is risk neutral. A rule of thumb which takes care of uncertainty, but which retains a standard investment calculation, would involve, according to (9.29) subtracting from the discounted value ΔR_{ig} the cost k. It is readily seen that this cost k

increases when the calculation rate decreases (Π_2 and V_2 acquire proportionately equally great positive shifts);
increases when future willingness to pay for recreational amenities increases in relation to the value of power (V_2 increases in relation to Π_2).

9.5 OPTION VALUES

As noted in the introduction to this chapter, if consumers behave as *homo economicus*, we would expect them to be willing to pay for the *option* to consume recreational services in the future. When the development of, say, watercourses which destroy nature is considered, option values should clearly be added to the development costs, thereby making developments less favourable. This is the case whether the consumers actually consume the recreational services in the future or not. The way we have discussed option values so far clearly indicates that these values are connected with future events and hence a dynamic framework is called for, such as the one used in the preceding section. Actually, the magnitude k defined in (9.29) can be considered to reflect the option values of increased future information. The reason is that k measures the difference in net income of a project calculated at expected values, ignoring learning over time in an uncertain and irreversible environment, and the same project calculated in a way which implies an optimal strategy with respect to learning in this same environment.

Instead of pursuing option values in a dynamic framework we shall deal with these values in a static model with the purpose of showing how these values are related to other concepts used in decisions under uncertainty. Our exposition follows Hanemann (1984).

Consider a representative individual equipped with a utility function which depends on the two arguments: income, and whether a recreational area is destroyed or not.

Thus,

$$U = U(r,d); \quad U'_r > 0, U''_r < 0 \tag{9.30}$$

where U is utility, r is income and d represents the state of the recreational area.

y denotes gross income and k denotes costs in the development of say, water courses for hydropower production or the opportunity cost of preservation. The choices facing the individual can be described as follows:

	Decisions	Costs	Net income
Preserve nature	$d = 0$	k_0	$r_0 = y - k_0$
Destroy nature	$d = 1$	k_1	$r_1 = y - k_1$

Uncertainty may enter this problem in different ways. We will assume that the net incomes are uncertain, dependent on what states of the world that occur. Let s denote state s and r_{0s}, r_{1s} the corresponding net incomes. State s occurs with the probability πs $(s = 1, ..., S)$.

Applying the expected utility theorem the individual will choose preservation, $d = 0$, if

$$\Delta U = \sum_s \Pi_s \, U(r_{0s}, 0) - \sum_s \Pi_s \, U(r_{1s}, 1) \geq 0 \tag{9.31}$$

The option value, OPV, is defined implicitly by

$$\sum_s \Pi_s \, U(r_{0s} - \text{OPV}, 0) = \sum_s \Pi_s \, U(r_s, 1) \qquad (9.32)$$

and says how much the individual is willing to pay for preservation and thereby avoiding the irreversible action of destroying nature in an uncertain environment.

Combining (9.31) and (9.32) we get

$$\Delta U = \sum_s \Pi_s \, [\, U(r_{0s}, 0) - U(r_{0s} - \text{OPV}, 0)\,] \qquad (9.33)$$

Since $U'r > 0$, then

$$\text{sign } \Delta U = \text{sign OPV} \qquad (9.34)$$

Thus, the same decision is reached with respect to development whether one decides on the basis of differences in expected utilities (ΔU) or on the basis of the money measure OPV.

The reason why we are doing this mathematical exercise is the fact that it might be easier to reveal OPV from questioning the individuals than obtaining estimates of the utility functions.

A widely used measure to characterise uncertain incomes is the certainty equivalence of the uncertain incomes. Let $\hat{r}_0$ and $\hat{r}_1$ denote the certainty equivalences of the two uncertain incomes r_{0s} and r_{1s}. $\hat{r}_0$ and $\hat{r}_1$ are defined as follows:

(i) $\qquad U(\hat{r}_0) = \sum_s \Pi_s \, U(r_{0s}, 0) \qquad (9.35)$

(ii) $\qquad U(\hat{r}_1) = \sum_s \Pi_s \, (U(r_{1s}, 1) $

Applying the basic criterion (9.31) and using (9.35) we then get

$$\Delta U \geq 0 \quad \text{if } U(\hat{r}_0) \geq U(\hat{r}_1) \qquad (9.36)$$

or since $U'_r > 0$

$$\Delta U \geq 0 \quad \text{if } \Delta CE \equiv \hat{r}_0 - \hat{r}_1 > 0 \tag{9.37}$$

where ΔCE is the differences of the certainty equivalents.

Thus, another equivalence is established which together with (9.34) gives us the following equivalences:

$$\text{sign } \Delta U = \text{sign OPV} = \text{sign } \Delta CE \tag{9.38}$$

Whether one should try to measure OPV or ΔCE seems to be a matter of taste of the observer. They are closely related and the same technique of questioning or revelation of preferences seems to be required.

Some have suggested another measure which they argue is easier to estimate from individual preferencs.

Let C_s denote the willingness to pay for preservation conditional on the occurrence of state s. Thus,

$$U(r_{0s} - C_s, 0) = U(r_{1s}, 1) \tag{9.39}$$

Let COPV denotes the *conditional* option value defined by

$$\text{COPV} = \sum_s \Pi_s C_s \tag{9.40}$$

From (9.39) we clearly see that

$$C_s = r_{1s} - r_{0s} \tag{9.41}$$

such that

$$\text{COPV} = \sum \Pi_s (r_{1s} - r_{0s}) = \sum \Pi_s r_{1s} - \sum \Pi_s r_{0s} \equiv \bar{r}_1 - \bar{r}_0 \tag{9.42}$$

where $\bar{r}_1$ and $\bar{r}_0$ are the expected value of the uncertain incomes r_{1s} and r_{0s}.

The distinction between OPV and COPV thus is that OPV is in utility terms or derived from a concept equivalent to utility terms whereas COPV is not related to utility terms at all. It is simply the differences in expected values of the two incomes. It

223

therefore comes as no surprise that decisions based on COPV are equivalent to decisions based on OPV only in the case of a linear utility function which means risk-neutrality. This can be demonstrated by using risk premiums p_0 and p_1, respectively, and they are defined as the difference between the certain equivalence of the income and expected income, that is

(i) $p_0 = \hat{r}_0 - \hat{r}_0$ (9.43)

(ii) $p_1 = \hat{r}_1 - \hat{r}_1$

From the definition of ΔCE and COPV we then get

$$\Delta CE = \text{COPV} + p_1 - p_0 \qquad (9.44)$$

Suppose $p_1 > p_0$ which means that the variance of incomes r_{1s} is higher than the variance of incomes r_{0s}, or shorter, r_{1s} is more risky than r_{0s}. Then,

$$\Delta CE > \text{COPV}$$

We could clearly then have the case that while COPV < 0, $\Delta CE \geq 0$. Hence, there is no equivalence present in basing decisions on conditional option values and expected utilities — or unconditional option values (OPV). Although it might be easier to estimate COPV it could introduce a bias in the decisions if this is done.

REFERENCES AND FURTHER READING

Arrow, K.J. and A.C. Fisher (1974): 'Environmental preservation, uncertainty and irreversibility', *Quarterly Journal of Economics*, 88, 312-19

Fisher, A.C. and J.V. Krutilla (1985): 'Economics of nature preservation' in *Handbook in natural resources and energy economics*, vol. 1, (ed.) A.V. Kneese, North Holland, Amsterdam

Hanemann, M.W. (1984): 'On reconciling different concepts of option values', Californian Agricultural Experiment Station, Giannini Foundation of Agricultural Economics

Henry, C. (1974): 'Investment decisions under uncertainty: the irreversibility effect', *American Economic Review*

Krutilla, J.V. (1967): 'Conservation reconsidered', *American Economic Review*, 64, 1006-12

Krutilla, J. et al. (1972): *Natural environments*, Johns Hopkins Press, Baltimore

McCounel, K.E. (1985): 'The economics of outdoor recreation', in *Handbook in natural resources and energy economics*, vol. 1, (ed.) A.V. Kneese, North Holland, Amsterdam

Part III

Natural Resources

10

The Exploitation of Non-renewable Resources

10.1 INTRODUCTION

Towards the end of the last century and round about the turn of the present century there was a public debate, just as heated as the present one, revolving round the question of the exhaustion of the world's raw material resources by the industrial countries. The discussion abated somewhat when the First World War broke out. The inter-war years, with their low utilisation of capacity, feeble economic growth, and mass unemployment, moreover, provided no basis for renewing this discussion. The Second World War, too, provided the industrial countries with problems other than the exhaustion of resources and the protection of the natural environment. The period after the Second World War, and up to the beginning of the 1970s, was marked by steady economic growth in most countries; at the very least any interruption in the growth rate proved of far more modest scope than during the inter-war years. This steady post-war growth can be explained in part by the high rates of investment, technical and organisational progress and the more methodical political management of many countries' economies. Not until the 1970s was the debate on the exhaustion of resources once again resumed, and given increasing urgency in view of the sharp rise in the price of oil in the 1970s.

With the aid of data from the USA for the period 1870-1957 Barnett and Morse test the hypothesis of increasing shortage of raw materials and the attendant danger of the exhaustion of resources. They construct time-series showing the development in prices for extractive products in relation to prices for non-extractive products. Their most important empirical results do not support the hypothesis of an increasing shortage of resources; on the contrary, the opposite appears to have taken place. Unit costs in extractive production have either remained constant or fallen as compared with the costs of other products. One of the aims of this chapter is to show how, on the basis of economic theory, we can foretell the development in the prices of extractive products.

In sections 10.2-10.4 we shall discuss what might determine the price development of products based on non-renewable resources. By non-renewable resources we mean resources such as oil, gas and minerals. These are often called exhaustive, but this might also happen for renewable resources such as forests and fish.

In section 10.5 we shall examine factors that might explain the price development discovered by Barnett and Morse. Furthermore, we shall deal with a number of circumstances which make it imperative to exercise caution in drawing a 'comforting' conclusion, and that many have been tempted to draw, namely, that results of the kind arrived at by Barnett and Morse show that we are not moving in the direction of increasing shortage of natural resources.

The discussion of shortages of natural resources has long figured in economic theory. Thomas Malthus and David Ricardo were both concerned with the limits natural resources would place as a bar to continued growth. Ricardo's point of departure was that access to natural resources was almost unlimited, though the quality might vary. Gradually, as qualitatively poorer and poorer natural resources were utilised, production costs in the extractive industries would undergo a real rise. In this way, in fact, the natural basis would set a limit to growth. The results arrived at by Barnett and Morse lead one to question Ricardo's conclusions; as already mentioned, we shall investigate this problem more closely at a later stage in this chapter.

The analytical contribution that was to provide the foundation to studies of how the community should, and how different marketing forms would, exploit non-renewable resources was submitted by Harold Hotelling in 1931. The problem he dealt with was how perfect competition and monopoly would utilise non-renewable resources. An important approach, furthermore, in Hotelling's case was to demonstrate the need for regulating decisions to deplete with the aid of price mechanisms, instead of the politically popular and consequently generally employed method, namely, direct regulation, bans and mandatory orders.

10.2 EXPLOITATION OF A NON-RENEWABLE RESOURCE: AN INTUITIVE APPROACH

10.2.1 The perfect competition solution

The individual producer in a free-competition economy takes the price of the product as given. The product we have in mind could be oil, gas, or a mineral. The individual producer has available one or more oil or gas reservoirs or one or several mines. His production is at all times so small in relation to the total supply that his particular supply does not influence the market price. The producer may be assumed to organise his production in such a way that discounted profit is as great as possible. Adjustment can then be described as follows.

The producer is at all times faced with a problem of investment. Where would he have the best chance of making money? By extracting the product and putting the profit out on interest in a bank or by allowing the product to remain in the reservoir or mine and in this way cashing in on a rise in price? An optimal production plan is obviously characterised by a balance in the yield of these two forms of investment. We shall demonstrate more explicitly precisely what this adjustment is. Let $P(t)$ be the price of the product and $C(t)$ be the extraction cost per unit: $r(t)$ is the interest the producer gets on a bank deposit in a credit market. The price, the unit cost, and the interest are all deflated so that they are real magnitudes.

Let us assume that the producer extracts a unit of the product at point of time t. The net revenue at point of time t is then $P(t) - C(t)$. This net income is invested at interest in the capital market. During the period from t to $t + \Delta t$ the producer has in this way achieved a yield of

$$[P(t) - C(t)][1 + r(t)\,\Delta t] \tag{10.1}$$

Let us assume that the producer allows this unit to remain in the reservoir/mine for a period from t to $t + \Delta t$. We then calculate

231

that both the price and the unit cost may change. By postponing production by a unit during the period t to $t + \Delta t$ the producer thereby achieves a profit of:

$$P(t+\Delta t) - C(t+\Delta t) \tag{10.2}$$

Provided the yields produced by the two methods of investment are equally big, the producer is indifferent where he is to make his investment. This means that a successful production plan is characterised by the fact that $(10.1) = (10.2)$

$$\frac{P(t+\Delta t) - C(t+\Delta t) - (P(t) - C(t))}{\Delta t}$$

$$= r(t)(P(t) - C(t))$$

Now, letting $\Delta t \to 0$, we have directly from the definition of the derivative:

$$\frac{\dot{P}(t) - \dot{C}(t)}{P(t) - C(t)} = r(t) \tag{10.3}$$

(10.3) tells us that the net price $P(t) - C(t)$ per unit produced will increase at a rate equal to the interest rate. $\dot{P}(t)$ and $\dot{C}(t)$ are the derivatives of the functions $P(t)$ and $C(t)$ with respect to t, respectively.

We shall for the moment assume that the unit cost $C(t)$ is independent of the magnitude of production and the stocks of resources. Hence, $C(t)$ is included in $P(t)$. We shall also ignore the difference in costs between one producer and another. A free competition economy is characterised by the fact that all producers are faced with the same price. (10.3), or more briefly (10.4), therefore describes an equilibrium in a market with perfect competition:

$$\frac{\dot{P}(t)}{P(t)} = r(t) \tag{10.4}$$

Eq. (10.4) is derived from the balance between the two profit flows in (10.1) and (10.2). Eq. (10.4) can also be derived from a balance between capital stocks. This can most easily be seen if we assume that the interest rate $r(t)$ is constant over time. The value today of a mining product is P_0. If this is extracted today and the profits are invested in financial markets, the financial capital will be $P_0 e^{rt}$ at point of time t. If instead the producer allows the product to remain in the mine/reservoir, the capital at point of time t will be $P(t)$. In our perfect model world the producer will be indifferent with respect to where this capital is invested. His capital equilibrium will thereby be characterised by:

$$P(t) = P_0 e^{rt} \qquad (10.5)$$

From (10.5), (10.4) follows when r is constant.

We have now almost reached the conditions that characterise the equilibrium in a free competition market for the extraction of products from non-renewable resources.

The demand function of the market for the product, written on the price form, is

$$P(t) = P(x(t)), \qquad (10.6)$$

where P on the right side is a function symbol and where $x(t)$ is the total production at point of time t.

We shall subsequently return to a more detailed description of the demand function. The price development described in (10.4) or (10.5) is an equilibrium price development. This means that the price development in (10.4) or, where applicable, (10.5) must be the same as in (10.6). If we use (10.5) we get

$$P_0 e^{rt} = P(x(t)) \qquad (10.7)$$

Since r is given in the financial market, (10.7) will determine the production at every point of time when P_0 is determind. What is it that determines P_0?

In order to answer this question we shall have to look more closely at the demand function. The product we have in mind most frequently has 'perfect' substitutes. This means that an upper limit $\bar{P}$ exists, which is such that when the price P rises above this limit, the demand for the product will diminish and switch over to the substitute. This substitute in turn may have a substitute, and so on. Concrete examples might be petrol from oil. The substitute could be wood alcohol. The substitute for wood alcohol could be sawdust, and so on. The reason why we now use petrol and not wood-alcohol as fuel for cars is that at the moment petrol is cheaper. The unit cost of methylated spirits, however, is hovering in the background. Methylated spirits in this case acts as a so-called 'backstop technology' in relation to petrol. Is there any need at all to make use of wood-alcohol? Yes, we shall have to do so, since there is a limited amount of oil in the world. From (10.5), or, where applicable, (10.4), we see that it is a question of time before the price $P(t)$ rises to the 'backstop' price $\bar{P}$. In a situation of this kind the amount of resources left in the mine/reservoir will have no value. However, intuitively it seems reasonable in a perfect competitive world with perfect foresight that when the price $P(t)$ reaches the level of $\bar{P}$ the resources will be exhausted. We shall call the point of time for exhaustion T. If we use the formula (10.5) we then get

$$\bar{P} = P(T) = P_0 \, e^{rt} \tag{10.8}$$

Since $\bar{P}$ is a 'backstop' price, it follows from the demand function when unit extraction cost is constant that $x(T) = 0$.

However, we have not yet determined P_0. Admittedly, we have a new equation, but we also have a new unknown, namely, the lifetime T. But since T is the point of time of exhaustion, the stocks of resources at point of time T, $S(T)$, must be zero. Let S_0 indicate the known initial stocks of resources; we then have the following condition

$$S_0 = \int_0^T x(t)\mathrm{d}t \tag{10.9}$$

Eq. (10.5) (where applicable (10.4)), (10.6), (10.8) and (10.9) thereby determine the unknowns

$$P(t), x(t) \quad \text{for all } 0 \le t \le T$$

and T, as functions of the exogenous parameters S_0, r (where applicable $r(t)$) and $\bar{P}$. Thus we see that the price in a perfect competitive world with perfect foresight will adjust itself in such a way that the profit from 'digging up' the product and investing the net income in a bank will be at all times exactly equal the capital gains involved in allowing the product to remain in the mine/reservoir. The initial price and thereby the initial production will adjust itself in such a way that this, together with the price/production development for the subsequent points of time, will result in the resource being exhausted during a period of finite time. If this is to occur, the product must have a substitute. If the product has no substitute, it means that the 'backstop' price $\bar{P}$ is infinitely great. We are then dealing with a product we cannot do without. In that case the lifetime of the resource will be infinite, and the amount of the resource will asymptotically approach zero. We do not know the existence of a product of this nature, and we shall ignore this eventuality in what follows.

If we use formula (10.5), we shall see that the price of the product is bound to rise monotonically until it reaches the 'backstop' or the 'substitute' price $\bar{P}$. Since the demand curve is normally a falling one, this means that production in the case of free competition would fall monotonically towards zero. The price and production developments are illustrated in Figure (10.1).

Figure 10.1: Price and production developments in a free competition market for a product from a non-renewable natural resource

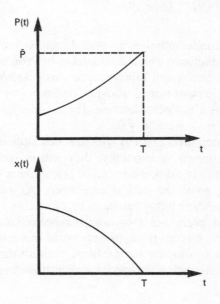

235

10.2.2 Is the competitive solution optimal?

Hotelling's answer to this question was yes. The explanation is as follows. Let $U(x)$ represent the utility for the representative individual, here assumed to be equal to the social utility of the amount of good x. We shall use partial reasoning, as we ignore all other goods. If we allow the community's objective to be maximisation of the discounted utility, given the physical limitations that exist and the properties of the utility function, and only these, we get the following problem:

$$\max \int_0^T U(x(t)) \, e^{-rt} dt \quad \text{with respect to } x(t) \text{ and } T \quad (10.10)$$

given

(i) $\quad \dot{S}(t) = -x(t)$

(ii) $\quad S(T) = S_0 - \int_0^T x(t) dt \geq 0$

(iii) $\quad x(t) \geq 0$

(iv) $\quad U'(0) = \bar{U}' > 0$

In the economic optimum we shall therefore find the production plan and a lifetime (which *a priori* may be infinitely great) which makes the discounted utility as great as possible. We shall use the same discount rate as above. Condition (iv) tells us that the product has a perfect substitute. If this were not the case, then $\lim U'(x) = \infty$ when $x \to 0$.

If we compared (10.10) with the description in (10.2.1) of the free competitive economy, they will in fact coincide. The free competitive market consists of price-takers which maximise the utility, given the budget conditions. As we are here considering the same utility function, in a free competitive economy one will at every point of time maximise $U(x)$, given $Px =$ income available for the consumption of this product.

In order to simplify the problem, without the generality and the conclusion being changed, we shall therefore assume a constant budget share. The consumers' adjustment is thus described by means of the condition

$$U'(x) = P$$

Remembering from the free-competition section above that the producer maximises discounted profit, given physical constraints which of necessity cannot be different from (i) to (iii), we shall see that in the free-competitive economy the following expression is maximised:

$$\int_0^T e^{-rt} \left(\int_0^{x(t)} U'(y) \, dy \right) dt = \int_0^T e^{-rt} U(x(t)) \, dt$$

The social optimum problem and the optimum problem implied by the free competitive economy are therefore the same optimum problem. Note that in assuming the representative individual we ignore possible problems of distribution.

If, nevertheless, we argue that the free competition solution does not provide the economic optimum solution, we shall have to look for indirect effects and distribution considerations. We shall single out the following three circumstances:

(1) Extractive activity involves residuals (which are a necessary but not sufficient condition for pollution) and other environmental effects, often of a lasting character. As a general rule, a free competitive economy will not take these indirect effects into account on its own.

(2) Several independent producers operate on the same resource base. A case in point would be two oil producers producing oil from the same reservoir. What one producer can extract per unit of time, and over a given period of time, will then depend on what the other producer extracts and has extracted in the past. Since we are here speaking of indirect effects, as a general rule the agents in a free competitive economy will not take these into account in their adjustment.

(3) Economic conservation motive. This motive may either be expressed in such a way that a positive utility is associated at every point of time with the stocks of certain raw materials, or that a positive utility is associated with

237

the possession of stocks of certain raw materials at the termination of given planning periods. The positive utility can be explained on the basis of the uncertainty with regard to which substitutes will be available in future for stocks of raw materials which have been exhausted. The objection that can be raised to a point of view of this kind is that further *potential choice* in a society would not automatically increase because stocks of certain raw materials increase. A society's future potential choice must be associated with the *level* and *composition* of *total capital* in that country. If we consider one community, its total capital will consist of four components:

(1) Natural capital
(2) Real capital
(3) Human capital
(4) Net claims on other countries

The conservation of certain specific raw materials is no guarantee that future potential choice will be the widest possible. A reduction in the rate of extraction of certain raw materials may make future stocks of, for example, real and human capital less. We shall return to this in greater detail in the last section of the next chapter.

(4) Public intervention in a competitive economy may also be based on other distribution considerations. A positive discount rate attaches greater weight to events in the present and the immediate future than to events in the distant future. In *The Economics of Welfare*, (1920), A.C. Pigou states:

'But there is a wide agreement that the State should protect the interests of the future in some degree against the effects of our irrational discounting and of our preference for ourselves over our descendants. The whole movement for 'conservation' in the United States is based on this conviction. It is the clear duty of Government, which is the trustee for unborn generations as well as for its present citizens, to watch over, and, if need be by legislative enactment to defend, the exhaustible natural resources of the country from rash and reckless spoliation.'

The interest rate implicit in a free competitive economy is partly determined by the agents' preference for the future. What Pigou's point of view involves is that the State's appreciation of the present as against the future differs from that of the agents. Individual time preferences do not harmonise with preference in a collective body. This paternalist attitude has to a large extent been accepted in most countries, in so far as in most of them the position is that the state has been entrusted with a great deal of the responsibility for overall savings.

10.2.3 The monopoly case

As above, we shall let the extraction costs be independent of production and stocks of resources, and let $P(t)$ be the net price at the point of time t. The monopolist, in his adjustment, must take into account the fact that he is faced with the demand functions of a market. The extra income of the monopolist at point of time t as a result of extracting an extra unit of resource is thus (somewhat imprecisely) the marginal income

$$\frac{\mathrm{d}(P(x) \cdot x)}{\mathrm{d}x} \equiv \frac{\mathrm{d}a(x)}{\mathrm{d}x} \equiv a'(x) \tag{10.11}$$

where, in other words, $a'(x)$ is the monopolist's marginal income.

If this net marginal income is invested at interest in a bank in period $t + \Delta t$, the monopolist will achieve a return of

$$a'(x(t))(1 + r(t)\Delta t) \tag{10.12}$$

If the monopolist allows this unit to remain in the mine/reservoir during the period t to $t + \Delta t$, he will achieve a profit of

$$[a'(x(t)) + \dot{a}'(x(t))\Delta t] \tag{10.13}$$

Just as in the free competitive case, the two profits will equal one another if the producer has chosen an optimal production or savings plan. We then get:

239

$$\frac{\dot{a}'\,(x(t))}{a'\,(x(t))} = r(t) \tag{10.14}$$

In the monopolist case, in other words, marginal income will increase at a rate equal to the interest rate.

From the definition of marginal income it follows that

$$a'\,(x) = P(x)\,(1 + \check{P}(x))$$

where $\check{P}(x)$ is the elasticity of P with respect to x.

If we call the function, $1 + P(x)$, $\varepsilon(x)$, we get

$$\frac{\dot{a}'(x)}{a'(x)} = \frac{\dot{P}}{P} + \frac{\dot{\varepsilon}}{\varepsilon} = \frac{\dot{P}}{P} + \frac{\varepsilon'(x)}{\varepsilon(x)} \cdot \dot{x}$$

But as $\dot{P} = P'(x) \cdot \dot{x}$ the last expression can be written

$$\frac{a'(x)\dot{P}}{a'(x)P}\,(1 + \frac{\check{\varepsilon}(x)}{\check{P}(x)})$$

where $\check{\varepsilon}(x)$ and $\check{P}(x)$ are the elasticities of ε and P with respect to x. But,

$$\check{\varepsilon}(x) = \frac{\check{P}\ \check{\check{P}}}{1 + \check{P}}$$

so that

$$\frac{\dot{a}'(x)}{\dot{a}'(x)} = \frac{\dot{P}}{P}(1 + \frac{\check{\check{P}}}{1 + \check{P}}) \tag{10.15}$$

If we insert (10.15) in (10.14), and set $r(t)$ equal to a constant r, we get

$$\frac{\dot{P}}{P} = \frac{1 + \check{P}}{1 + \check{P} + \check{\check{P}}} \cdot r \tag{10.16}$$

(10.16) is the optimal price development in a monopoly corresponding to the price development (10.4) in a competitive

economy. Before comparing price and production developments within the two market forms, we shall, as in the competitive case, have all prices, production amounts, and lifetimes
determined. (10.16) together with the demand function

$$P(t) = P(x(t)) \tag{10.17}$$

the 'backstop' price demand,

$$P(T) = \bar{P} \tag{10.18}$$

and the resource condition

$$S_0 = \int_0^T x(t)\,dt \tag{10.19}$$

determine the unknowns $x(t)$, $P(t)$ and T as functions of the
exogenous variables S_0, $\bar{P}$ abnd r (or, where applicable, $r(t)$).

We have then tacitly assumed that the 'backstop' technology
is in the hands of a competitive group of producers. They are
ready to start production as soon as the monopolist has
exhausted his production potentials.

We shall now analyse the optimal price development given in
(10.16). The following two questions are of interest:

(1) Can the price in the monopolist case fall during any
period?
(2) What is the price development like in the monopolist
case compared to the competitive solution?

In order to answer these questions we first note that the monopolist selects the production development and the lifetime of his
resource which makes the discounted profit as great as possible,
given as opposed to the free competition case, the demand
function and given the constraint associated with the exhaustion
of resources. This means that the monopolist's maximisation
problem corresponding with the reasoning which led to (10.16)
is:

$$\max \int_0^T P(t)x(t)\,e^{-rt}dt \quad \text{with respect to } x(t) \text{ and } T \quad (10.20)$$

241

given

$$P(t) = P(x(t)) \text{ and where } P = P(0)$$

$$\dot{S}(t) = -x(t)$$

$$S(T) = S_0 - \int_0^T x(t)\,dt \geq 0$$

$$x(t) \geq 0$$

An analytical and numerical solution of (10.20) will be shown in the next section. In this section it should suffice to note in the first place that the necessary condition that solves the problem will entail the requirement that the marginal income is non-negative. This means that

$$a'(x) = P(1 + \check{P}) \geq 0$$

Since $P > 0$, this means that $(1 + \check{P}) \geq 0$.

The sufficiency condition involves the requirement that the function which is maximised is strictly concave. This means that the function $a(x)$ is strictly concave. We easily obtain that this means that

$$a''(x) = P \cdot x \cdot \check{P}[1 + \check{P} + \check{\check{P}}] < 0.$$

Since the demand function is falling, $\check{P} < 0$. This means that

$$1 + \check{P} + \check{\check{P}} > 0.$$

From (10.16) we see that in this case we are bound to get

$$\dot{P}/P \geq 0$$

for all values of x and consequently for all values of t. Just as in the competitive case the price must be non-decreasing. Since the demand curve is falling, production must be non-increasing. Normally the price would rise monotonically to reach the 'backstop' price, and production would fall monotonically and be equal to zero at the point of exhaustion.

If we call the rate in price increase in the competitive case $\dot{P}_F / P_F$ and in the monopoly case $\dot{P}_M / P_M$, we observe from (10.5) and (10.6) that

242

$$\frac{\dot{P}_M}{P_M} - \frac{\dot{P}_F}{P_F} = \frac{r}{1 + \check{P} + \dot{P}} (-\check{P})$$

This means that

$$\frac{\dot{P}_M}{P_M} > \frac{\dot{P}_F}{P_F} \quad \text{when } \check{P}(x) < 0$$

$$\frac{\dot{P}_M}{P_M} = \frac{\dot{P}_F}{P_F} \quad \text{when } \check{P}(x) = 0 \qquad (10.21)$$

$$\frac{\dot{P}_M}{P_M} < \frac{\dot{P}_F}{P_F} \quad \text{when } \check{P}(x) > 0$$

If we consider the actual demand function more closely we discover that

$$\check{P}(x) = x \frac{P''(x)}{P'(x)} + (1 - \check{P}(x)) \qquad (10.22)$$

Here, P' and $\check{P}$ are both negative, since the demand curve is falling. From (10.21) and (10.22) it then follows that

(i) When the demand curve everywhere is linear or concave, the monopoly price will always rise less strongly than the free competition price, see Figures 10.2 and 10.5.

(ii) When the demand curve is everywhere strictly convex or convex over certain sections of the demand curve, it is uncertain how the monopoly price will rise in relation to the free competition price. We can then not exclude the possibility that the monopoly price in certain periods will rise more markedly than the free competition price; see Figures 10.6 and 10.7.

When the demand curve is strictly convex the price flexibility is greatest when the demand for the product is small and therefore the price is close to the 'backstop' price. This is probably the reason why this case is of less practical interest, but in principle

243

Figure 10.2: Linear demand curve

Figure 10.3: Corresponding price developments, monopoly and free competition

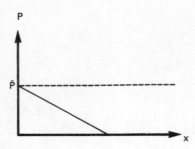

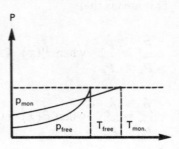

Figure 10.4: Concave demand curve

Figure 10.5: Corresponding price development

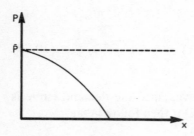

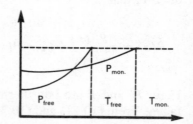

Figure 10.6: Mixed convex-concave demand curve

Figure 10.7: Possible, but not necessary price developments

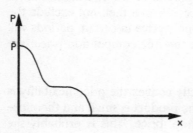

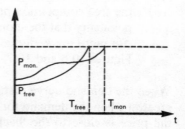

it cannot be excluded. One must, however, be able to conclude that the commonest case is for the demand curve to be more elastic as the price approximates to the price of the competitive substitute..

The case where $\dot{P}(x) = 0$ for all values of x is not possible, since this means that no final 'backstop' price exists. $\dot{P}(x)$ can, however, be zero along certain parts of the demand curve. A possible and reasonable variant is for the demand curve to be very slightly elastic when the price is low, subsequently becoming more elastic in an intermediate interval, maybe so that it is logarithmically linear in parts of this interval, finally becoming more elastic again when the price approximates to the 'backstop' price; see Figures 10.6 and 10.7.

10.3 THE CASE OF A LINEAR DEMAND FUNCTION

Monopoly

$$\max_{x(t),T} \int_0^T P(t)\, x(t)\, e^{-rt} dt$$

given

(a) $P(t) = P(x(t)) = \bar{P} - \beta\, x(t)$ (10.23)

(b) $\dot{S}(t) = -x(t)$

(c) $S(T) = S_0 - \int_0^T x(t) dt \geq 0$

(d) $x(t) \geq 0$

The monopolist maximises discounted profit. We note that $a(x) = Px$ is the monopolist's total income per unit of time. His marginal income is then $a'(x) = P(1 + \dot{P})$. Constant unit costs have been taken care of in the demand function, which, according to (a), is linear. The 'backstop' price is equal to $\bar{P}$. We can interpret $\bar{P}$ to reflect the difference in unit costs in the production of the substitute good and the good we are considering. A falling demand curve requires that $\beta > 0$. (b) tells us that the mine/reservoir is tapped with an intensity $x(t)$ equal to production per time-unit; (c) tells us that the final stocks are to be

245

non-negative; and (d) tells us that production at any point of time must be non-negative.

Problem (10.23) is a problem within the theory of optimal control. Necessary conditions that solve this problem are

(i) $a'(x(t)) = \bar{P} - 2\beta\, x(t) \leq \Lambda_0\, e^{rt}\ (= \Lambda_0 e^{rt}$ or $x(t) = 0)$
(ii) $P(t) = \bar{P} - \beta\, x(t)$ for all t
(iii) $(P(T) - \Lambda_0 e^{rT})\, x(T) = 0$

$$ \text{(10.24)} $$

(iv) $S(T) = S_0 - \int_0^T x(t)\mathrm{d}t \geq 0$

(v) $\Lambda_0 S(T) = 0$
(vi) $\Lambda_0 \geq 0$.

Λ_0 is a shadow price associated with stocks $S(t)$. In our case it is constant, since stocks $S(t)$ enter neither into the objective function nor into the differential equation (10.23) (b).

Conclusion 1. Production is equal to nought at the point of time of exhaustion

$$ x(\mathrm{T}) = 0. \tag{10.25} $$

Proof: Assume $x(T) > 0$. From (iii) it then follows that $P(T) = \Lambda_0 e^{rT}$

From (i) and (ii) we see that $P(T) - \beta\, x(T) = \Lambda_0 e^{rT}$.
For this reason we must get $x(T) = (P(T) - \Lambda_0 e^{rT})\, 1/\beta = 0$.
Contradiction.

Conclusion 2. The final price is equal to the 'backstop' price

$$ P(T) = \bar{P} \tag{10.26} $$

Proof: Follows from (10.25) and (ii).

Conclusion 3. The entire resource is exhausted

$$ S(T) = 0 \tag{10.27} $$

Proof: Assume that $\Lambda_0 = 0$. From Conclusion 1 and (i) it follows that at T we must get $\bar{P} \leq 0$.

Now $\bar{P} > 0$. This is consequently a contradiction. From (v) it then follows that $S(T) = 0$.

Conclusion 4. The shadow price Λ_0 is equal to the discounted value of the final price (equal to the 'backstop' price)

$$P(T) = \Lambda_0 e^{rT}. \tag{10.28}$$

Proof: From Conclusion 1, 2 and (i) it follows that $P(T) \le \Lambda_0 e^{rT}$. (10.28) fits.

Conclusion 5. Production is strictly positive during the entire period of usage. Marginal income increases at a rate equal to the discount rate

$$x(t) > 0 \quad \text{for } t\varepsilon\ [0, T) \tag{10.29}$$

and therefore from (i)

$$a'(x) = \bar{P} - 2\beta x(t) = \Lambda_0 e^{rt} \quad \text{for } t\varepsilon\ [0,\ T) \tag{10.30}$$

Proof: Assume $x(t') = 0$ for a $t'\varepsilon\ [0,\ T)$. Then we must get $\bar{P} \le \Lambda_0 e^{rt'}$.
But since $\bar{P} = \Lambda_0 e^{rT}$ according to (10.28) and (10.26) we then get $e^{rt} \le e^{rt'}$. In that case we must have $t' > T$; a contradiction.
It will be seen directly from (10.30) that we get

$$\frac{\dot{a}'(x)}{a'(x)} = r \tag{10.31}$$

We now have the following equations that describe the optimum of the monopoly and determine the unknown Λ_0, T, and $x(t)$:

$$x(T) = 0 \tag{10.25}$$

$$S_0 = \int_0^T x(t)\,dt \tag{10.27}$$

Figure 10.8: Development of production

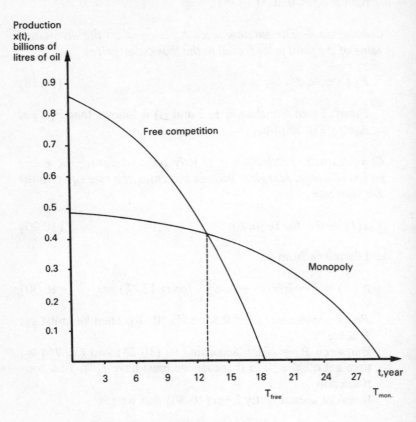

$$\bar{P} - 2\beta x(t) = \Lambda_0 e^{rt} \quad \text{for all } 0 \le t \le T \tag{10.30}$$

When we know $x(t)$ we can find $P(t)$ from the demand function (10.24), (ii) and $S(t)$ from the formula

$$S(t) = S_0 - \int_0^t x(\tau) d\tau.$$

If we solve these we get the following recursive system:
T determined in:

$$T + \frac{1}{r}(1 - e^{-rT}) = \frac{2\beta S_0}{\bar{P}} \tag{10.32}$$

248

Figure 10.9: Development of prices

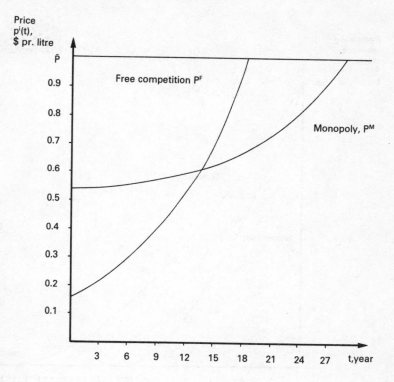

then $x(t)$ is determined in

$$x(t) = \frac{\bar{P}}{2\beta}\left(1 - e^{-r(T-t)}\right) \tag{10.33}$$

and hence Λ_0 is determined by:

$$\Lambda_0 = \bar{P}\,e^{-rT} \tag{10.34}$$

The time development for $x(t)$, $P(t)$ and $S(t)$ is given in Figures 10.8-10.10.

From (10.32) we can see that increased initial stocks of resources, increased S_0, or a steeper demand curve, increased β

249

Figure 10.10: Resource depletion

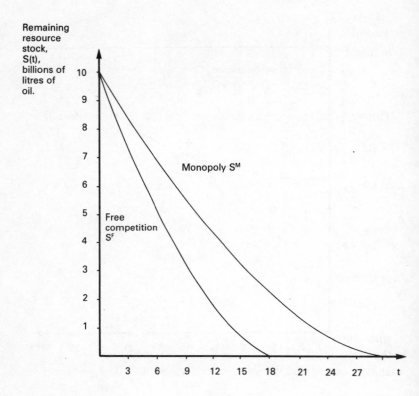

(consumers with the lowest willingness to pay are eliminated) and a reduced 'backstop' price have the same effect on the lifetime and result in its extension. An increased discount rate reduces the lifetime.

The competitive solution (= social optimum)

$$\max \int_0^T P(t)x(t)\,e^{-rt}dt \quad \text{with respect to } x(t) \text{ and } T \quad (10.35)$$

given

$$\dot{S}(t) = -x(t)$$

$$S(T) = S_0 - \int_0^T x(t)dt \geq 0$$

$$x(t) \geq 0$$

In addition we shall demand market equilibrium

$$P(t) = P(x(t)) = \bar{P} - \beta x(t) \tag{10.36}$$

The necessary conditions for solving this problem are:

    (i) $P(t) \leq \Lambda_0 e^{rt}$ for $0 \leq t \leq T$ $(= \Lambda_0 e^{rt}$ or $x(t) = 0)$

    (ii) $[P(T) - \Lambda_0 e^{rT}] \, x(T) = 0$

    (iii) $S(T) = S_0 - \int_0^T x(t)dt \geq 0 \tag{10.37}$

    (iv) $\Lambda_0 \, S(T) = 0$

    (v) $\Lambda_0 \geq 0$

where Λ_0, as before, is a constant shadow price associated with the stock $S(t)$. Without going through the proofs, which will be analogous to those above, our conclusions will be as follows:

Conclusion 1. Production is equal to zero at the terminal point of time

$$x(T) = 0 \tag{10.38}$$

Conclusion 2. The final price is equal to the 'backstop' price

$$P(T) = \bar{P} \tag{10.39}$$

Conclusion 3. The entire resource is exhausted

$$S(T) = S_0 - \int_0^T x(t)dt = 0 \tag{10.40}$$

251

Conclusion 4. The shadow price Λ_0 (equal to the initial price P_0) is equal to the discounted value of the final price

$$\Lambda_0 = P(T) e^{-rT} \tag{10.41}$$

Conclusion 5. Prices grow at a rate equal to the rate of interest

$$P(t) = \Lambda_0 e^{rt} \quad (\Rightarrow \frac{\dot{P}}{P} = r) \tag{10.42}$$

We now get the following equations for solving the unknowns Λ_0, T, and $x(t)$.

$$x(T) = 0 \tag{10.38}$$

$$S_0 = \int_0^T x(t) dt \tag{10.40}$$

From (10.36) and (10.42):

$$\bar{P} - \beta x(t) = \Lambda_0 e^{rt}$$

If we know $x(t)$ we can find $P(t)$ from the demand function (10.36), and $S(t)$ from the formula

$$S(t) = S_0 - \int_0^t x(\tau) d\tau.$$

If we solve the equations we get the following recursive system:

$$T + \frac{1}{r}(1 - e^{-rT}) = \frac{\beta S_0}{\bar{P}} \tag{10.43}$$

$$x(t) = \frac{\bar{P}}{\beta}(1 - e^{-r(T-t)}) \qquad\qquad (10.44)$$

$$\Lambda_0 = \bar{P}e^{-rT} \qquad\qquad (10.45)$$

The time development for $x(t)$, $P(t)$ and $S(t)$ is given in Figures 10.8–10.10. (In comparison of (10.33) and (10.44) it might be tempting to say that for every t monopoly production is half the free competitive production. This is wrong, since the lifetime T is different in the two solutions.)

In Figures 10.8-10.10 we have reproduced production, price, and resource developments when the parameters S_0, β, $\bar{P}$, and r have the following values:

$$S_0 \quad = \quad 10 \text{ (billion litres of oil)}$$
$$\bar{P} \quad = \quad 1 \text{ ($1 per litre)}$$
$$r \quad = \quad 0.1 \text{ (10 per cent discount rate)}$$

We then get:

the lifetime in the free competitive case is about 18 years;
the lifetime in the monopoly case is about 29 years.

Thus a monopoly results in the resource being exhausted later than is economically optimal.

During the first 13 years or so production in free competition is greater than in a monopoly. The monopolist thus restricts production unduly during the first phase of the lifetime of, for example, the reservoir, and during this phase he demands too high a price. In the last phase the price is lower than would be optimal. We see that a monopoly results in too weak a rise in the price of the product. The monopolist, as already mentioned, exhausts the resource too late. We see from Figure 10.10 that the remaining stocks of resources in a monopoly are always greater than would be economically optimal.

Finally, we shall look at the effects on the lifetime of changes in the rate of interest in the two market systems (see Table 10.1). We can see that in free competition (equal to the social optimum) the lifetime increases relatively more strongly with a reduction in the discount rate than in the monopoly. From the two columns we can see that the monopoly with a sufficiently high requirement for rate of return in excess of the market rate

Table 10.1

Social rate of discount	Lifetime in free competition	Lifetime in a monopoly
8 per cent	20	32
10 per cent	18	30
12 per cent	17	28
20 per cent	15	25

(equal to the optimal rate) can behave as though it were free competition and thus in accordance with the socially optimal solution.

10.4 STOCKS OF RESOURCES AND PRODUCTION INFLUENCE COSTS

We shall restrict ourselves to discussing the free competition case. The problem is now:

$$\max \int_0^T [P(t)\, x(t) - C(x(\mathrm{t}),\, S(t))]\, \mathrm{e}^{-rt}\mathrm{d}t \qquad (10.46)$$

with respect to T and $x(t)$

given

(i) $\dot{S}(t) = -x(t)$

(ii) $S(0) = S_0$

(iii) $S(T) = S_0 - \int_0^T x(t)\mathrm{d}t \geq 0$

(iv) $x(t) \geq 0$

In addition we require the market to clear at every point of time:

$$P(t) = \bar{P} - \beta\, x(t) \qquad (10.47)$$

What is new in this problem is the cost function $C(x,S)$. We assume that

$$C'_x > 0 \quad C''_x \geq 0 \tag{10.48}$$
$$C'_S < 0$$

(10.48) implies that marginal costs are no longer constant, but rise as production rises. Furthermore, we presuppose that as the stocks of resources are exhausted, so extraction costs rise. A cost function satisfying (10.48) is

$$C(x,S) = \frac{1}{2}\left(\frac{x}{S}\right)^2 \tag{10.49}$$

This cost function, furthermore, is of such a kind that

$$\min_{S \to 0} C'_x(x,S) = +\infty \tag{10.50}$$

Marginal extraction costs tend towards infinity when stocks of resources tend towards zero. The necessary conditions for solving problem (10.46) are:

(a) $\quad P(t) - C'_x(x(t), S(t)) = q(t)$

(b) $\quad \dot{q}(t) = rq(t) + C'_S(x(t), S(t))$ \hfill (10.51)

(c) $\quad e^{-rT}q(T)\, S(T) = 0$

(d) $\quad e^{-rT}\{[P(T)\, x(T) - C(x(T), S(T))]$

$\qquad - q(T)\, x(T)\} = 0$

In addition we have side conditions (i)–(iv) and the market equilibrium condition (10.47).

From (10.51) we see that the difference between price and marginal extraction costs should be equal to the shadow price $q(t)$ associated with the stock of resource $S(t)$. When stocks S are not present in the cost function, we see from (10.51) (b) that this shadow price will increase with the interest r. This, too, was the result we arrived at in the previous section and which was first discovered by Harold Hotelling. Since stocks of resources are now integrated in the cost function it is no longer such a simple matter to decide how the net price will develop.

We note first of all that since (10.50) applies, exhausting resources completely will never actually occur. With zero stocks of resources the marginal extraction costs will be infinitely great, and it will long since have proved profitable for consumers to change over to the substitute with the constant 'backstop' price $\bar{P}$. This means that

$$S(T) > 0. \tag{10.52}$$

From (c) it then follows that

$$q(T) = 0 \tag{10.53}$$

and therefore according to (10.51) (a) and (d)

$$P(T) = C'_x(x(T), S(T))$$
$$P(T) \cdot x(T) = C(x(T), S(T)).$$

These two are only fulfilled simultaneously for

$$x(T) = 0 \tag{10.54}$$

From (10.47) we see that

$$P(T) = \bar{P} \tag{10.55}$$

Thus, as previously, production at the terminal point of time must be equal to zero and the production price equal to the 'backstop' price.

With the aid of a phase diagram analysis we shall finally discuss the development of the central variables. From (10.46) (i), (10.47) and (10.49) we can write (10.51) (a):

$$\dot{S} = \frac{q - \bar{P}}{\beta + \dfrac{1}{S^2}} \qquad\qquad (10.51)(a')$$

A corresponding approach makes it possible for (10.51) (b) to be written

$$\dot{q} = rq - \frac{(q - \bar{P})^2}{(\beta + \dfrac{1}{S^2})^2} \, S^3 \qquad\qquad (10.51)(b')$$

In Figure 10.11 we show the functions $q = f_1(S)$ and $q = f_2(S)$ which have been arrived at setting $\dot{S} = 0$ and $\dot{q} = 0$ in (10.51) (a) and (b), respectively.

Figure 10.11 illustrates a situation in which the depletion resources starts from an initial stock S_0. In accordance with Hotelling's simple case where marginal extraction costs are independent of stocks of resources, the shadow price q and consequently the net price $P - C'_x$ will rise to start with. In this

Figure 10.11: Development of stocks of resources and shadow price equal to the price minus marginal extraction costs when stocks of resources affect the cost function

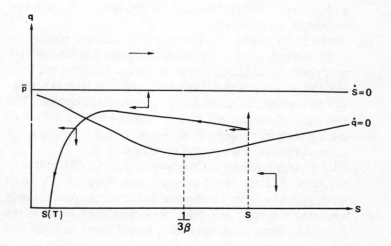

257

phase, by straightforward arithmetic, we discover as previously that production, $x(t)$, is bound to fall and the price $P(t)$ rise. In the last phase the net price is bound to fall, since on reaching the point of time of exhaustion it must be equal to zero. In this last phase the net price, in fact, behaves differently from what it doeṣ in the simple Hotelling case. The resource continues to be exhausted. It may prove difficult to find an unambiguous answer analytically to the question of whether production will rise or fall. But since $x(T) = 0$ and consequently $P(T) = P$, the price $P(t)$ is bound to rise in the last phase in the direction of the 'backstop' price. Thereby production, too, will fall in this last phase of the economic lifetime of the resource.

10.5 OBSERVATIONS ON BARNETT AND MORSE'S RESULTS IN FALLING RELATIVE PRICES OF EXTRACTIVE PRODUCTS

The above model shows uniquely that net prices for extractive products will rise in relation to other prices. This is a result which is not in accordance with the results discovered by Barnett and Morse. However, the following reasons may explain why they arrive at different results:

(1) In the production of extractive goods there may be more marked technical progress than in other sectors. As we have mentioned, this is a conclusion drawn by Barnett and Morse and others. This hypothesis, however, has not been tested sufficiently against data.
(2) Research on substitutes, with a view to technical production improvements so that the 'backstop' price can be lowered, may result in producers of raw materials reducing the price in order to weaken this R & D activity. The R & D activity will have a smaller scope than is economically optimal. The raw material market will then belong to a different form of market than free competition. Sooner or later, however, the price of extractive goods will then have to rise.
(3) The strategy outlined in (2) above indicates a different market form than the two forms we have discussed. A cartel strategy, and one in which possible intruders are constantly excluded through 'limit-price' strategies, may result in the price not rising as sharply as it would have done in free

competition or in a monopoly of the type we have discussed.

(4) In Norway, Norwegian power projects are not selected in such a way that costs are minimised. In Norwegian power supplies there is a slight tendency for expensive projects to be selected in preference to cheap ones. The result will be that unit costs decline actually over a period of time. At one time, however, they are bound to rise sharply. A similar occurrence may have been the case in the production of the goods analysed by Barnett and Morse.

(5) Prices can be fixed in such a way that historical unit costs are to be covered. If this is to prove to be the case, several prices must exist for the same good in the same market. The law of indifference has been broken. It is difficult to see how this can occur in the case of goods that can be stored and in this way bought up by speculators and in a market where there is freedom of trade. The sale of electricity, however, is an example of an item that cannot be stored. In Norway the cost-price principle has been the predominant system of price fixing. To a certain extent this is also the case with the production and sale of crude oil in the USA.

(6) The production of certain extractive products is more capital-intensive than production in other industries. As real capital has been accumulated, so the return on capital has dropped. Capital costs have thereby decreased in relation to other costs, and consquently, too, unit costs in extractive activities.

(7) Insufficient data is available, nor are the statistical methods used by Barnett and Morse good enough, for the results they arrived at to be accepted as fully substantiated.

All in all, the conclusion must be that even though there *may* be something in the explanation that technical progress can rescue us from a future increasing shortage of raw materials, nevertheless there are so many counter-arguments that can be raised that caution should be exercised in drawing such a comfortable conclusion.

(8) The exhaustion of resources may have an effect on the discount rate r. In the first place oil is an important input factor used in most sectors of the economy. In some cases,

259

too, it would be correct to say that real capital and oil are complementary factors in the technical sense. A drop in supplies of oil, which is an implication in the models above, will thus result in the marginal productivity of capital dropping for a given amount of capital. If, in addition, real capital, too, rises over time, this will accentuate the fall in the marginal productivity of capital. Technical progress may counteract the drop. The marginal productivity of capital can under certain reasonable assumptions be shown to be equal to the rate of interest. A possible falling rate of interest, whether it is generated by a slump in supplies of raw material or capital accumulation, will slow down the price increase for raw materials; see, for example, (10.4) in the free competitive case. In a monopoly situation we must expect the monopolist to take into consideration the fact that the rate of interest depends on extraction of a raw material, but in this case, too, it may be shown that a monopolist will have the incentive to reduce the rate of price increase. Changes in the rate of interest will thus result in the depletion of production not proceeding as quickly as suggested in the partial models above. The lifetime of the resource will be extended.

The same levelling of the price curve will also be the result if we consider that income invested in the capital market may influence the yield provided by financial investment. This in particular will be the case if we are dealing with a raw material such as oil, which is used everywhere in the world economy and consequently produces large incomes for relatively few producers. An important feature of many of the oil-producing countries is, furthermore, that the savings rate is abnormally great. One of the reasons for this is that several of the oil-producing countries in OPEC have a very uneven income distribution. Marginal consumption propensity is therefore small. Since a large proportion of the income is invested in financial markets, this will directly influence the rate of interest that will help to determine the price path. The marked savings rate among the oil-producing countries can in this way contribute to levelling the price path we arrived at in the partial models above.

10.6 A SIMPLE MACRO-ECONOMIC CONSIDERATION OF THE RELATION BETWEEN THE EXHAUSTION OF RESOURCES AND INVESTMENT ACTIVITY

The model is:

$$C_t = \beta_c K_{c,t} \qquad \text{Production function in the consumer goods sector} \qquad (10.56)$$

$$I_t = \beta_I K_{I,t} \qquad \left.\begin{array}{l}\\ \text{production function in the investment goods sector}\end{array}\right\} \qquad (10.57)$$

$$R_t = v\,I_t \qquad\qquad\qquad\qquad\qquad\qquad (10.58)$$

$$S_t = S_{t-1} - R_t \qquad \text{resource depletion} \qquad (10.59)$$

$$K_{c,t} = K_{c,t-1} + I_{c,t-1} \qquad\qquad\qquad (10.60)$$

$$\left.\begin{array}{l}\\ \end{array}\right\} \text{capital accumulation}$$

$$K_{I,t} = K_{I,t-1} + I_{I,t-1} \qquad\qquad\qquad (10.61)$$

$$\Lambda_{c,t} = I_{c,t}/I_t \qquad \text{allocation of total} \qquad (10.62)$$

$$\left.\begin{array}{l}\\ \end{array}\right\} \text{investment to two}$$

$$\Lambda_{I,t} = I_{I,t}/I_t \qquad \text{sectors} \qquad (10.63)$$

$$\Lambda_{I,t} = 1 - \Lambda_{c,t} \qquad \text{adding up investment shares} \qquad (10.64)$$

$$Y_t = C_t + I_t \qquad \text{Total supply equal to demand} \qquad (10.65)$$

C_t, I_t, Y_t and R_t are respectively consumption, investment, gross natural product, and extraction of raw materials per unit of time. $K_{c,t}$ and $K_{I,t}$ are stocks of real capital in the consumer goods and investment goods sectors. $I_{c,t}$ and $I_{I,t}$ are the corresponding gross investments. $\Lambda_{c,t}$ and $\Lambda_{I,t}$ are allocation shares. $S(t)$ is the stock of the resource.

The only input factor in the production of consumer goods is real capital. In the investment goods sector raw materials and capital are used. We shall ignore the inputs used in the extraction of resources, that is, of raw materials.

When $\Lambda_{I,t}$ is given, the model is determined. By straightforward calculation we discover

(i) $I_t = (1 + \beta_I \Lambda_I) I_{t-1}$

When I_0 is known and when $\beta_I \Lambda_I$ are given figures, all subsequent investments are thereby determined. We have presupposed that $\Lambda_{I,t}$ is a constant, that is, that $\Lambda_{I,t} = \Lambda_I$ for all values of t.

(ii) $C_t = C_{t-1} + \beta_c \Lambda_c I_{t-1}$

When β_c, Λ_c and C_0 are known, we can deduce C_t from (i) and (ii).

(iii) $Y_t = C_t + I_t$

Y_t determined by (i) and (ii).

(iv) $S_t = S_0 - v \sum_{\tau=0}^{t} I_\tau$

When S_0 is known, we can deduce S_t from (i) and (iv). We carry out a numerical reference solution by substituting

I_0 $= 0.2$, $C_0 = 0.8$ $(Y_0 = 1.0)$

β_c $= 0.2$

β_I $= 0.4$

v $= 0.35$

and Λ_I $= 0.5$ for all values of t.

Parameter values agree reasonably with Norwegian conditions as of the beginning of the 1980s. We use the model to discuss the following problem:

What will be the effects in the future on investment, consumption, and gross natural product of introducing as a requirement that the remaining stocks of resources should be 65 per cent greater in 15 years' time?

This demand involves the following, namely, that

$[S_0 - S^{15}]$ alternative path $= 0.35 \, [S^0 - S^{15}]$ reference path

This means that Λ_I alternative path $= \hat{\Lambda}_I$ is the unknown.

$$\sum_{\tau=0}^{15} (1+0.4 \, \hat{\Lambda}_I)^\tau = 0.35 \sum_{\tau=0}^{15} (1+0.5 \cdot 0.5)^\tau$$

This is a polynomial of order 15. By iteration we get

$$\Lambda_I = 0.2$$

The conclusion will then be that if the share allocated to investment in the capital good sector falls from 0.5 to 0.2 the remaining stocks of natural resources will be increased by 65 per cent in 15 years' time.

The consequences for capital investment, consumption, and the national product are shown in Figures 10.12-10.14. From Figure 10.15 we can see that in 5 years' time the remaining stock of resources will be greater on the 'environmental protection (conservation) curve' than in the reference curve. Real capital in the economy, on the other hand, will be lower in 15

Figure 10.12: The development of investment, reference path and 'conservation path'

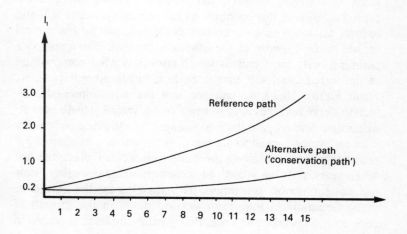

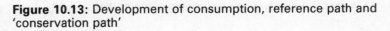

Figure 10.13: Development of consumption, reference path and 'conservation path'

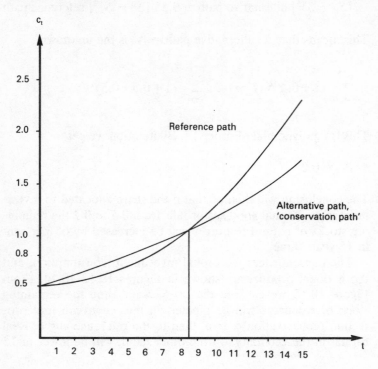

years' time in the 'conservation path': see Figure 10.12. Total capital consists in this economy of the two components real and natural capital. The conservation path increases future natural capital at the expense of the other component. The question a planning body must endeavour to answer is what composition of the total capital will ensure the best future potential choice. From Figure 10.13 we can see that the environmental protection curve involves a more pronounced initial growth of consumption. Annual growth of consumption in the first eight years is on an average equal to 4.6 per cent per annum, as against 2.8 per cent per annum along the reference curve. During the last seven years or so the growth of consumption is far weaker along the environmental protection path than along the reference path. Generally, a long-term slowing down in the growth of consumption is taken as a sign that greater emphasis is placed in

264

Figure 10.14: Gross national product, reference path and 'conservation path'

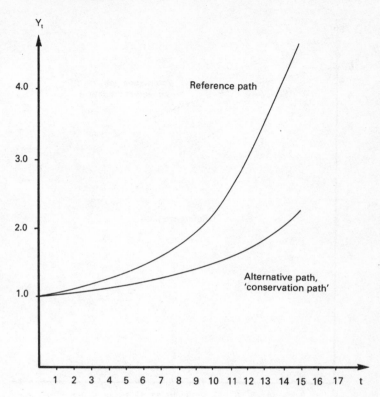

the economy on the *present time* as opposed to the future. This might appear paradoxical within the framework of our model, since the environmental conservation resulted in higher remaining stocks of resource in 15 years' time. But we should remember that consumption involves not only consumption of 'Nature'. If the exhaustion of natural resources for production purposes takes place at the expense of consumption of Nature, our calculated consumption development will underestimate the level and growth of *total* consumption along the environmental protection path. Nevertheless, we should be able to conclude that greater emphasis on environmental protection will result in an initial period with more vigorous growth of consumption, subsequently followed by a period with a lower growth rate of consumption. In other words, an increase in momentary enjoyment

265

Figure 10.15: Remaining stock of resources, reference path and 'conservation path'

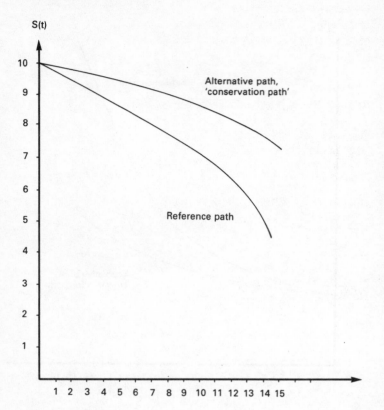

of consumer goods is a result of a greater emphasis on the natural capital in the future.

An important tacit assumption to this conclusion is that production capacity is throughout fully exploited. The portion of the investment goods production which is not supplied to the investment goods sector itself is used throughout in the consumer goods sector.

Another assumption that has also been used, but which does not play an equally important role, is that established production capacity in one sector cannot be used in another (C_0 and I_0 are not choice variables and the capital K_c cannot be transferred to the stocks K_I).

Ex ante investment goods can, in other words, be used anywhere, but not after they have been installed as capital in one of the sectors. If one assume that capital is used in the extraction of raw materials, the conclusion above, as far as the development of consumption and investment is concerned, will merely be reinforced.

REFERENCES AND FURTHER READING

Dasgupta, P. and G.M. Heal (1974): 'The optimal depletion of exhaustible resources', *Review of Economic Studies, Symposium on the economics of exhaustible resources*, 3-28

Dasgupta, P. and G.M. Heal (1979): *Economic theory and exhaustible resources*, Cambridge University Press, Cambridge

Gilbert, R.J. (1978): 'Dominant firm pricing in a market for an exhaustible resource', *Bell Journal of Economy*, 9, 305-95

Herfindahl, O.C. (1967): 'Depletion and economic theory' in *Resource economics, selected works*, Johns Hopkins Press, Baltimore

Hoel, M. (1978): 'Resource extraction under some alternative market structures' in the series *Mathematical systems in economics*, no. 39, Verlag Anton Hain

Hotelling, H. (1931): 'The economics of exhaustible resources', *Journal of Political Economy*, 39, 137-75

Stiglitz, J.E. (1976): 'Monopoly and the rate of extraction of exhaustible resources', *American Economic Review*, 66, 655-61

Sweeney, J.L. (1977): 'Economics of depletable resources: market forces and intertemporal bias', *Review of Economic Studies*, 44, 125-42

11

Commercial Fishing

In this chapter we shall analyse the cropping of biological resources that are self-reproductive. As a typical example we have chosen fishing, although our analysis could equally well apply to such resources as forests and herds of animals. An important distinction exists between resources to which a person may have property rights, for example, a forest, and resources where this does not apply (for example, pelagic or ocean fishing). In the last-mentioned case, too, a conflict may easily arise between social and private exploitation of resources on account of the '$1/n$ effect', see Chapter 5, section 5.2

11.1 INTRODUCTORY REMARKS ON REGULATED VERSUS UNREGULATED FISHING

The social costs involved in commercial fishing consist of two components: the first are the private costs associated with fishing, such as wage expenses, the cost of fuel, etc. For the moment we shall assume that the total private costs are proportional to the quantity fished. This means that private unit costs are constant. The other costs are opportunity costs that are not registered in ordinary markets. The point is that a fish may be used in other ways than would be the case if it ended up on the quay. If the fish escapes being caught, it may grow larger by next year. It may also reproduce, thus ensuring that there will be more fish in future. The discounted value of the gains created by a kilo of uncaught fish we shall call the price of one kilo of uncaught fish. This price will depend on future demands for fish, on technological development within commercial fisheries, the importance future is given in social investment calculations, the size of stocks of fish, etc. The fewer fish are left in the area, the greater the value an uncaught fish will generally have. The price of a kilo of uncaught fish cannot be observed in the market; it must be calculated by a body possessing all the information mentioned above.

The price of fish as delivered on the quay is observable; it depends on a number of factors, such as the total catch of fish. We shall assume in order to simplify that the price is given and constant. The characteristic feature of socially optimal fishing will then be this: that social marginal costs are equal to the price of the fish supplied on the quay. If less than this quantity is fished, the price for a kilo of fish delivered on the quay will be greater than would be the cost of catching another kilo of fish. It would then be worth while catching at least one extra kilo. If more than the right quantity were fished, the cost of fishing the last kilo would be greater than the price. In our simple model we shall therefore assume that it would be prudent to organise fishing in such a way that the social marginal costs are equal to the price of a kilo of fish delivered on the quay. We can express this adjustment in the following way, remembering our assumption with regard to constant unit costs:

Condition A

Price per kilo of fish caught and delivered at the quay — private unit costs = price per kilo of uncaught fish

Condition (A) tells us that the *net* price of a kilo of fish caught is equal to the price of an uncaught fish.

In most cases the private unit cost will be greater than zero. By way of example, let us assume that this is equal to 50p. Furthermore, let us assume that the adjustment above results in the price of fish, delivered at the quay, is £1. From condition (A) we see that the price of a kilo of uncaught fish is 50p. This means that 'a fish in the hand is worth two in the sea'. (Assuming, of course, that the fish is of the same type and weighs just as much.) We can see that a fish 'in the hand' will be worth more than a fish 'in the sea', provided the unit costs are greater than zero.

The characteristic feature of unorganised fishing will be that

269

the individual fisherman will act as though an uncaught fish is worth nothing. Briefly, we might say that so little weight is attached to the future profits of a kilo of uncaught fish that this profit disappears completely. 'Who knows whether I of all people will catch this fish next year?' It could be maintained that in that case all stocks of fish would have to be exterminated, which conflicts with actual observations. We do not intend to consider the problem of whether, in fact, entirely unorganised fishing has existed, but if that is the case, the following effect will damp the tendencies to exterminate the fish: in the first place the price of fish delivered at the quay may drop when supplies increases. This will reduce earnings in the fishing industry to the level where private costs will be exactly covered. In such circumstances there would be no recruitment to the fishing industry. In the second place it would be unrealistic to assume constant unit costs if fishing is to be greatly increased. When fish caught start to be few and far between, fuel expenses, among other things, will increase markedly. On account of increased costs recruitment to fishing will be regulated. One should therefore expect cyclical movements over time in the number of boats participating in the unregulated fishery, in the total catch and in the stock. These expectations are for instance confirmed to a large extent in the Icelandic herring fishery after the Second World War, cf. section 11.4 below.

Even though unregulated fishing will not result in the extermination of fish, it will nevertheless never be the best solution. The socially best form of fishing will be characterised by condition (A). There are several ways of regulating fishing. One way is to introduce a charge levied on every kilo of fish caught equal to the price of a kilo of uncaught fish. Another method is to allocate catching quotas to independent fishermen. A third method is to entrust fishing to a centrally directed fishing authority. No matter what method is used, it will prove necessary to introduce the factual circumstance represented by the expression 'the price per kilo of uncaught fish'.

It will be difficult to put the regulation of fishing into practice if the fish are in international waters. The extension of fishing limits, international agreements, etc. will in such cases provide potential ways of regulating fishing. Inevitably, dynamic game situations will have to be discussed.

The formula above is naturally a highly simplified model of

complicated realities; the case might arise, for example, in which at a particular point in time we have a stock of fish that is in the process of being exterminated. If the net price that can be obtained for a kilo of fish supplied at the quay, when only one kilo is supplied, is less than the price for a kilo of uncaught fish, a complete ban on fishing will be introduced. This ban on fishing will last until stocks of fish have recuperated to a sufficient level for the net price for the marginal kilo of fish delivered at the quay to be greater than or at least equal to the price per kilo of uncaught fish.

The opposite extreme can also, in principle, occur. From the point of view of fishing biology stocks of fish may be too great. An uncaught fish will then not only be worth nothing, but it will have a negative price. The fisherman will be paid extra for every fish landed. This maximal fishing will continue until stocks once again have been sufficiently reduced for the price of a kilo of uncaught fish exactly to equal the net price of a kilo of fish delivered at the quay. In the intermediate case condition (A) will characterise the fishing that is economically most efficient.

If we compare stationary, unregulated fishing with regulated fishing, then in the last case the quantity will be smallest. In regulating fishing it will be important to obtain an estimate of the price of a kilo of uncaught fish. It will then be necessary at all times not only to be in possession of information on the location of stocks of fish, numbers, age structure, etc. but also on future demand and technical developments in commercial fisheries.

11.2 LIST OF VARIABLES

The variables refer to point of time t. Reference to this has been excluded, apart from the variables it is important to consider for the time notation.

x_i = quantity fished in tons by fishing vessel i
v_i = vector of factors employed by vessel i
$n(t)$ = number of fishing vessels at point of time t
$Z(t)$ = stocks of fish at point of time t
p = price per unit of fish delivered at the quay
q = price per unit of uncaught fish ('shadow price')
w = price vector for input factors

$$c_i \quad = \text{total costs, boat } i$$
$$\pi_i \quad = \text{profits, boat } i$$

$$\dot{Z}(t) \;=\; \frac{\mathrm{d}Z(t)}{\mathrm{d}t}, \; \dot{n}(t) = \frac{\mathrm{d}n(t)}{\mathrm{d}t}$$

Z and n are stock variables. The other variables are flow magnitudes.

11.3 THE PRODUCTION FUNCTION IN FISHING AND INDIRECT EFFECTS IN FISHING

With the aid of inputs v_i boat i engages in fishing. The boat's catch of production of fish will be represented by the function

$$x_i = b_i(v_i, Z), \quad \frac{\partial b_i}{\partial v_i} > 0 \quad \frac{\partial b_i}{\partial Z} > 0, \quad b_i(v_i, 0) = 0 \quad (11.1)$$

The funtion is assumed to be a regular *ultra passum* production function in factors v_i. Fishing results will depend on the input of factors and how much fish is to be found in the area where fishing is taking place. The denser the shoals of fish, the easier it will be to catch them. The greater Z is, the less the required fishing input for catching a given quantity of fish.

A reasonable supplementary demand on production function given in (11.1) is that an upper capacity limit exists for x_i, given v_i. In other words, after a choice of boat and type of tackle has been made, there is an upper limit to how much fish can be caught and landed. Another reasonable requirement is that even before a choice of boat and tackle has been made, an upper capacity level exists (greater than the one mentioned above). These capacity limits will not be *explicitly* introduced into the analysis.

Boat i has control of inputs v_i, but not over stocks of fish Z. The fact that stocks of fish are an integral part of the production function for fishing vessel i means therefore that there are external effects in fishing.

Two other external effects might also have to be considered. In the first place the fishing tackle may be such that unduly

272

young fish are caught; this will affect future stocks of fish. The mesh size on a trawl can be regulated in such a way that the smallest and youngest fish escape. The mesh width on a trawl for this reason affects the catches made by another vessel at a later point of time. In the second place the fish may appear in such dense shoals that the result will be queues on the fishing grounds. Below we shall ignore both the division of fish into age categories and the possibility of queues occurring. The external effect we shall study is the stock effect given in (11.1).

11.4 THE MINIMISED COST FUNCTION IN FISHING

We arrive at this cost function by minimising the cost, given a production quantity:

$$\min_{v_i} \{ c_i = w^T v_i \} \quad \text{given } x_i = b_i(v_i, Z) \tag{11.2}$$

where T denotes that w is transposed.

We are assuming a regular minimum. It is then well known that the minimised cost can be expressed as a function of the output and factor prices. We shall continue to keep factor prices constant. This means that the cost for boat i can be formulated as follows:

$$c_i = c_i(x_i, Z), \quad c'_{ix} > 0, \quad c''_{ixx} > 0, \quad c'_{iZ} < 0, \quad c''_{ixZ} < 0 \tag{11.3}$$

The assumption of a regular *ultra passum* law in factors, v_i, ensures that average costs have a minimum for variation in product quantity where marginal costs are equal to average costs.

11.5 STOCKS OF FISH

In the absence of fishermen stocks of fish are assumed to develop according to one of Nature's growth relations of the *Volterra* type:

$$\dot{Z} = f(Z) \tag{11.4}$$

where $f(Z)$ has the following properties

273

$$f(0) = f(\bar{Z}) = 0$$

$$f'(Z) \gtreqless 0 \text{ according to } Z \lesseqgtr Z^m \tag{11.5}$$

$$f''(Z) < 0$$

A function fulfilling (11.5) is

$$f(Z) = Z(\bar{Z} - Z), \ Z^m = \bar{Z}/2 \tag{11.6}$$

If left to themselves, stocks of fish will grow in the direction of $\bar{Z}$. The development is illustrated in the phase diagram in Figure 11.1. An alternative and possibly more realistic reproduction function is that a lower reproduction level exists. When stocks fall below this limit the population is moving in the direction of extermination. The function we assume is, however, simpler to work with, and implies that the lower reproduction level is equal to zero.

Figure 11.1

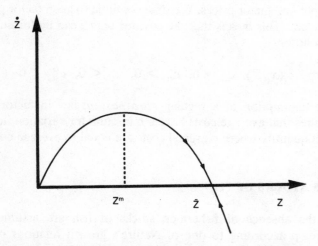

In Nature's stationary condition there are so many fish in the ocean that accretion has stopped. The growth rate when $Z = Z^m$ is the greatest growth rate Nature can produce. This growth rate $f(Z^m)$ may be called 'maximum sustainable yield', or MSY. By fishing we interfere in the development of stocks:

$$\dot{Z}(t) = f(Z(t)) - \sum_{i=1}^{n(t)} x_i(t) \qquad (11.7)$$

11.6 UNREGULATED FISHING

11.6.1 Introduction

Unregulated fishing consists of a number of fishermen $n(t)$ being 'let loose' on stocks of fish $Z(t)$ at every point of time t. There is no co-ordination between the fishermen: no individual fisherman is motivated to consider that by fishing 'today', he will be influencing fishing potentials of 'tomorrow'.

Each fishing boat is assumed to have adjusted its inputs as given in section 11.4. The quantity of fish caught and landed from each boat is assumed to be determined by maximising the fishing vessel's profit. The individual fisherman takes the price of fish delivered on the quay as given:

$$\max_{x_i} \{ \pi_i = px_i - c_i(x_i,Z) \} \quad \text{given } x_i \geq 0 \qquad (11.8)$$

which gives the necessary condition

$$p - c'_{ix}(x_i,Z) \leq 0 \ (=0 \text{ or } x_i = 0)$$

Below we shall only discuss an interior solution, so that

$$p = c'_{ix}(x_i,Z) \qquad (11.9)$$

applies.

We now introduce a simplification that does not notably reduce the generality of our reasoning. We assume that *all production functions b_i are alike*. Consequently, all cost functions will be alike, as well as the catch per boat. The subscripts i on the functions are therefore dropped.

To ensure that the model is complete, a few words will have to be said on what determines the price of fish delivered at the quay, and the number of fishing boats operating. We shall assume that p is given 'on the world market'. Catches of fish

275

from the fishing grounds we are considering do not affect the price of fish. It is assumed, furthermore, that the price p is constant over time.

The number of fishing boats entering and leaving this special fishing is determined by the profits obtained. With different cost functions it is reasonable to suppose that only boats with non-negative profits will operate. Since boats are now assumed to be alike, this entry/exit mechanism is simulated by introducing:

$$\dot{n}(t) = \delta \pi(t), \quad \delta > 0, \tag{11.10}$$

where

$$\pi(t) = px(t) - c(x(t), Z(t)). \tag{11.11}$$

It may be reasonable to operate with different values of δ, in so far as the profit per boat is positive or negative, so that the reaction coefficient is greater at exit than entry.

The consequences on development of stocks of fish of this unregulated fishing are

$$\dot{Z}(t) = f(Z(t)) - n(t)x)t), \quad Z(0) = Z_0 \tag{11.12}$$

All in all, we then have the following relations which constitute the model for unregulated fishing:

$$p = c'_x(x, Z) \tag{11.13}$$

$$\dot{n} = \delta \pi \tag{11.14}$$

$$\pi = px - c(x, Z) \tag{11.15}$$

$$\dot{Z} = f(Z) - nx \tag{11.16}$$

In this connection several problems can be raised, including the following:

(a) Will stocks of fish be exterminated?
(b) Will stocks of fish tend to be stable, stationary condition greater than nought?
(c) What characterises stationary stocks of fish; are stocks greater/equal or less than the stocks that produce MSY?

(d) Will the number of boats approach a stable, stationary level greater than zero?

11.6.2 Bio-economic equilibrium

If (b) and (d) occur, we say that we have bio-economic equilibrium. In this bio-equilibrium the following will apply:

$$p = c'_x(x, Z) \tag{11.13}$$

$$px = c(x, Z) \tag{11.17}$$

$$f(Z) = nx \tag{11.18}$$

From (11.13) and (11.17) we can see that x appears as a variable in both equations. Without Z in these two equations we should, now that p is given, have two equations to determine the one unknown, x. The solution in 'the usual case' is that p is not given, but determined in a market. In this way p depends on the quantity sold. The 'usual' mechanism is that firms adjust themselves as price takers, as given in (11.13) and without Z being involved. The firms are attracted to this branch, for example, as shown in (11.14). The price is depressed, until all pure profit has been eliminated and (11.17) occurs without Z being involved. In this 'ordinary' case, and where $g(\cdot)$ has no external effects, the unknowns are x and p. p depends on the total quantity nx. When p is known, n can therefore be calculated, since x is known. There is no reason why we could not have studied the same mechanism in our model, which in addition included negative, external effects: increased fishing on the part of the individual boats reduces stock of fish Z. In order to simplify the exposition we consider p as exogenously given and hence, x and Z are the only two endogenous variables in (11.13) and (11.17). Reduced stocks of fish result in increased costs. With the arrival of new fishing boats, involving a total increase in fishing, the negative external effects will also increase in their scope. The profits of the individual boat will drop. The following reasoning can illustrate this process: from (11.13) we see that x is a function of Z and p. From the second order conditions for profit maximum we find that x is an increasing function of Z and p. If we insert x as a function of

Z, p in the growth relation (11.16), for a given n, we can discuss how changes in the parameters n and p affect the development in Z. Figure 11.2 gives a diagrammatic idea of this.

For a given n, Z follows the unbroken line in the phase diagram at the bottom of the figure. For a bigger n the phase curve will shift inwards. For greater and greater n stock Z will tend

Figure 11.2

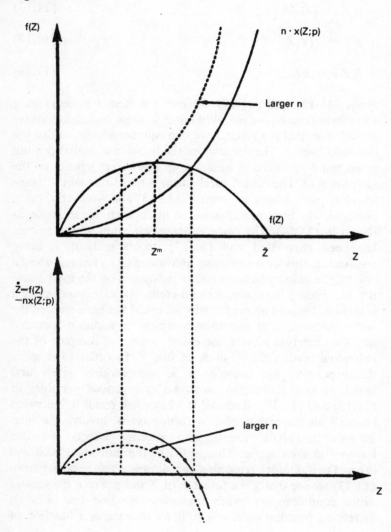

278

towards a lower and lower stationary value. From the cost function we see that for given x the costs will show increasing positive changes, as suggested in Figure 11.3. Initially, pure profits occur in this fishing. The price p is greater than average cost $\bar{c} = c/x$. More and more boats are attracted, until the pure profit is eliminated. This takes place when $p = \bar{c} = c'_x$. The last equality applies because the individual fishing boat must at all times be a profit-maximising price taker. The dashed curve suggests how $\bar{c} = c'_x$ gradually develops towards a point where $p = \bar{c} = c'_x$. We note that *catch per boat decreases constantly over time*. Figures 11.2 and 11.3 together indicate a process in which stocks of fish Z and the number of boats approach the stationary conditions Z^s and n^s. These are the conditions defined in the system of equations (11.13)–(11.18).

We shall now describe these stationary conditions in greater detail, as well as discussing whether n and Z will move in the direction of these stationary conditions.

Figure 11.3

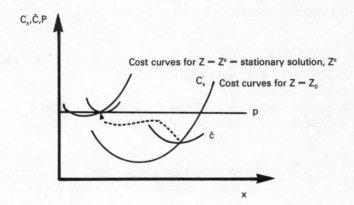

11.6.3 Characteristics of the stationary state

Let us consider the system of equations (11.13), (11.17) and (11.18). From (11.13) we find x as a function of p and Z, for example:

$$x^s = h(Z^s, p) \tag{11.19}$$

279

where the superscript s indicates stationary values.
We shall easily discover that

$$h'_z = -\frac{c''_{xz}}{c''_{xx}} > 0, \quad h'_p = c''_{xx} > 0 \tag{11.20}$$

Furthermore, we have

$$h(0,p) = 0. \tag{11.21}$$

If we insert (11.19) in (11.21) we get

$$Z^s = g(p) \tag{11.22}$$

It will readily be seen that $g' < 0$. For constant p, Z is a constant greater than 0. This is a stationary value for Z. For a higher value of the price p the corresponding stationary value will be lower.

If we insert (11.22) in (11.18) we shall find the corresponding equilibrium solution for n:

$$n^s = \frac{f(Z^s)}{h(Z^s,p)} = \frac{f(g(p))}{h(g(p),p)} \tag{11.23}$$

A possible stationary condition for Z is a Z^s so that $Z^s < Z^m$. This means that $f'(Z^s) > 0$. For total catch we then get

$$\frac{d(n^s x^s)}{dp} = \frac{df(g(p))}{dp} = f'(Z^s) \cdot g' < 0 \tag{11.24}$$

An increased product price will 'in the short term', that is, when n and Z are given, result in the total catch increasing. This is the usual conclusion which follows directly from (11.13) and from the second order conditions for profit maximum. 'In the long term', that is, in the future stationary condition n^s, Z^s, x^s, increased price will result in stocks of fish being lower, and in production, the quantity fished, being lower. From (11.23) we get

$$\frac{d n^s}{dp} = \frac{1}{h^2} \left(hf' g' - f(h'_z g' + h'_p) \right) \tag{11.25}$$

The sign is unknown. This will depend on the magnitude of the derivatives involves.

11.6.4 Movements in n and Z over time

We shall now discuss what the movement in Z and n can be over time. From (11.19), (11.14) and (11.15) it follows that:

$$\dot{n} = \delta \cdot [p\, h(Z,p) - c(h(Z,p), Z)] = \delta\, F(Z), \tag{11.26}$$

where the constant p is included in the functional form F.

(a) $F' = p\, h'_z - c'_x\, h'_z - c'_z = h'_z(p - c'_x) - c'_z = - c'_z > 0$

(b) $F(Z^s) = 0$ \hfill (11.27)

(a) follows on the basis of straightforward calculation and from (11.13). (b) follows from (11.19)–(11.22). This means that $\dot{n} = 0$ for $Z = Z^s$. From (11.27) it follows that for greater values of Z, F is greater, consequently F is positive and $\dot{n} > 0$. In other words

$$\dot{n} \underset{<}{\overset{>}{=}} 0 \text{ according to } Z \underset{<}{\overset{>}{=}} Z^s \tag{11.28}$$

From (11.16) and (11.19) we get

$$\dot{Z} = f(Z) - n(h(Z;p))$$

$$\dot{Z} = 0 \quad \text{for } n = \frac{f(Z)}{h(Z;p)} = n(Z) \tag{11.29}$$

$$\dot{Z} \underset{<}{\overset{>}{=}} 0 \text{ according to } n \underset{<}{\overset{>}{=}} n(Z) \tag{11.30}$$

Where the form of the function $n(Z)$ is concerned, we have:

$$n' = \frac{1}{h^2}(hf' - fh'_z). \tag{11.31}$$

For $Z > Z^m$, $f' < 0$ and consequently $n' < 0$. For $Z < Z^m$, $f' > 0$, and the sign for n' will depend on the magnitude included in the expression. It follows directly that the value for Z that gives the maximum for $n(Z)$ must be less than Z^m (since $f'(Z^m) = 0$ and $h'_z > 0$). For $Z = \bar{Z}$ we have $n(\bar{Z}) = 0$.

If we consider $n(Z)$ from $Z = \bar{Z}$ and towards the origin, the graph will curve down towards the origin if

$$hf' > fh'_z$$

One method of investigating this is to consider

$$\lim_{Z \to 0} \{ n(Z) = \frac{f(Z)}{h(Z;p)} \} \tag{11.32}$$

If this limit is zero, the graph for $n(Z)$ will curve down and approach the origin. As we have $f(0) = 0$, the limit will be zero if $h(0;p) > 0$. However, this is quite unreasonable, and we have assumed that the expression is equal to zero. We than get 2 a '0/0' expression. Applying l'Hôpital's rule we then consider

$$\lim_{Z \to 0} \frac{f'(Z)}{h'_z(Z;p)} \tag{11.33}$$

We have assumed that $f'(0) > 0$, so that the properties of the h function are once again decisive. It appears intuitively reasonable with

$$\lim_{Z \to 0} \{ h'_z(Z,p) = -\frac{c''_{xz}}{c''_{xx}} \} = \infty \tag{11.34}$$

Investigate, for example, the catch function $x = b^*(v) \cdot Z$. But it must be emphasised that examples of reasonable catch

282

functions can also be made which give nought as a limit value in the expression (11.34). We choose to assume that (11.34) applies.

We can now construct a phase diagram indicating development of n and Z. The direction of the arrows follows from (11.28) and (11.30), see Figure 11.4.

Figure 11.4

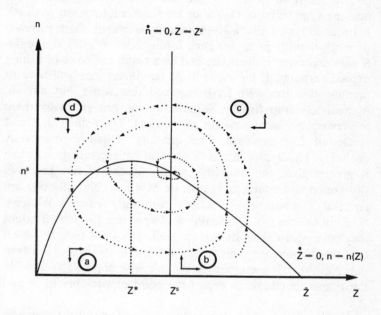

Z^* indicates the value of Z that maximises $n(Z)$. Z^m indicates the Z which maximises the biological growth relation $f(Z)$. $\bar{Z}$ is the saturation point for this relation. Z^s and n^s are stationary values for Z and n. For $Z = Z^s$ *the profit of fishing is equal to nought. For $Z < Z^s$* the profit is negative. For $Z > Z^s$ the profit is positive. The encircled letters divide the phase diagram into four areas.

The curve running in the direction of the arrows shows a possible development in the state variables Z and n. Let us assume that we start with 'large' stocks of fish, for example, $Z^m < Z_0 < \bar{Z}$ and a 'low' number of fishing boats, for example, such that (n_0, Z_0) is in the area (b). Fishing is abundant, but less is fished than Nature herself produces in the way of fish.

283

Stocks of fish, Z, increase. On the other hand, the rich fishing provides the fisheries with large profits. Several fishing boats join in order to cash in on this super-normal or pure profit. n increases. After a while so much fishing takes place that the quantity caught exceeds Nature's own production of the biomass. We have now reached the area (c). New fishing boats continue to join in. A pure profit still exists in the fishery. Stocks of fish, however, continue to decrease. This results in a positive shift in costs for the individual fishing boat. Fish become scarcer: more resources are required, for example, fuel, in order to ensure that a given quantity is caught. Costs increase so much that the profit for each fishing boat is equal to nought. Stocks continue to decrease, and as a result the costs of fishing exceed earnings. If fishing is to be continued, one will have to operate at a loss. We have assumed that some, but not all, immediately stop fishing. Some continue, but gradually there are fewer. We have now got to area (d). The fishing fleet and stocks of fish decrease. The quantity caught is constantly declining. Finally, so little is fished that net reproduction of fish is greater than the quantity caught. Stocks of fish increase. Fishermen still continue to operate at a loss. Not all costs are covered. The number of vessels is constantly reduced. We have now got to area (a). Gradually stocks recover to such an extent that once again all costs are covered. Stocks of fish, however, continue to increase. This results in lower costs involved in fishing. New vessels then join in. We are now in area (b). And so the rigmarole continues round the point of equilibrium n^s and Z^s.

Just how long we shall remain in areas (b) and (c) depends, *inter alia*, on the reaction parameter δ. The greater this is, the more speedily will new boats join in when pure profits are positive, and the more quickly will boats abandon this fishing when pure profits are negative.

11.6.5 Is the point of equilibrium n^s, Z^s stable?

At least two stability questions can be put:

Is the point of equilibrium (n^s, Z^s) locally asymptotically stable?

284

Is the point of equilibrium (n^s, Z^s) globally asymptotically stable?

By local asymptotic stability we mean that if the initial condition $n(0), Z(0)$ is sufficiently close to the stationary condition (n^s, Z^s), then $(n(t), Z(t))$ will tend towards the stationary state (n^s, Z^s).

By global asymptotic stability we mean that, no matter where the initial situation $n(0), Z(0)$ is to be found in the admissible area for (n, Z), $(n(t), Z(t))$ will tend towards the stationary condition (n^s, Z^s). Global stability implies local stability. The reverse implication, however, does not apply.

In order to investigate whether the solution $(n(t), Z(t))$ of the non-linear differential equations (11.26) and (11.29) are locally stable, we shall study this solution in a small area around (n^s, Z^s). We therefore develop the non-linear functions around this point with the aid of Taylor's formula. We then get the following two linear differential equations (remember that $F(Z^s) = f(Z^s) - n^s h(Z^s, p) = 0$, that n does not occur in $F(\cdot)$, and that (n^s, Z^s) are given numbers):

$$\dot{n} = \delta f'(Z^s)(Z - Z^s) \tag{11.35}$$

$$\dot{Z} = -h(Z^s, p)(n - n^s)$$

$$+ [f'(Z^s) - n^s h'_z(Z^s, p)] (Z - Z^s) \tag{11.36}$$

Around point (n^s, Z^s) the solutions $(n(t), Z(t))$ of (11.26) and (11.29) will behave more or less as indicated by the linear system (11.35) and (11.36). The conditions for *global* asymptotic stability of the solution of (11.35) and (11.36) and thereby the condition for *local* asymptotic stability of the solution of (11.26) and (11.29) is that the characteristic roots for the matrix

$$\begin{bmatrix} 0 & \delta F'(Z^s) \\ -h(Z^s, p) & f(Z^s) - n^s h'_z(Z^s, p) \end{bmatrix}$$

have negative real parts.

The characteristic equation for this matrix is provided by the equation

$$\begin{vmatrix} 0 - \Lambda & \delta\, F'(Z^s) \\ -h(Z^s, P) & f'(Z^s) - n^s\, h'_Z(Z^s, P) - \Lambda \end{vmatrix} = 0 \quad (11.37)$$

In connection with (11.28) we introduced the function $n(Z)$. This function applies to all n and Z values along the curve for $\dot{Z} = 0$, in particular for $n = n^s$ and $Z = Z^s$. We notice from the expression for n' given in (11.31) that at the point (n^s, Z^s) we have

$$f'(Z^s) - n^s h'_z(Z^s, P) = h(Z^s, P) \cdot \mathrm{n}'(Z^s) \quad (11.38)$$

By inserting this in the characteristic equation we shall find that this equation has the following roots

$$\Lambda = \frac{1}{2}\, h(Z^s, p)\, n'(Z^s)$$
$$\pm \sqrt{- \frac{h(Z^s, p)\, n'(Z^s)}{4} - \delta\, F'(Z^s) h(Z^s, p)} \quad (11.39)$$

Hence, $h(Z^s, P)$ and $\delta F'(Z^s)$ are both positive, so that the roots in the characteristic equation have negative real parts if and only if

$$n'(Z^s) < 0 \quad (11.40)$$

From (11.31) we then derive the result that this inequality is fulfilled for

$$Z^s > Z^*, \quad \text{since } n' = 0 \text{ for } Z = Z^*$$

The solution of the system of equations (11.26) and (11.29) in other words is locally asymptotically stable if and only if

$$Z^s > Z^*$$

In Figure 11.4 we have placed Z^s in such a way that this requirement is fulfilled. We note that in this case there is no need to ascertain the relationship of Z^s to Z^m.

From the definition of $n'(Z^s)$ we see that the requirement for stability can be transformed to give

$$El_Z f(Z)_{Z=Z^s} < El_Z (n^s x^s)_{Z=Z^s} \qquad (11.41)$$

In other words, a stationary state which is to be locally stable must fulfil the condition that in this state elasticity of Nature's reproduction with respect to stocks of fish is less than the elasticity of the profit-maximising fished quantity with respect to stocks of fish. Thus, local stable bio-economic equilibrium is characterised by the fact that, if stocks of fish had increased by, for example 1 per cent, then the quantity fished would have increased percentually more than nature's reproduction.

Global stability of the non-linear system is more difficult to investigate. In so far as local stability does not exist everywhere in the area of validity for Z^s, it follows that global stability does *not* exist.

The movements in the phase space shown by arrows in Figure 11.4 indicates a development converging on the point of equilibrium (n^s, Z^s). There is no guarantee that this will occur. If the initial situation $n(0)$, $Z(0)$ is sufficiently close to (n^0, Z^s), $(n(t), Z(t))$ will converge on the stationary state $(n^s n, Z^s)$. If the initial situation is a greater distance from the stationary state, we cannot exclude the possibility that we shall gradually, in cyclic movements, move further and further away from the stationary state. In principle we cannot exclude the possibility that unregulated fishing results in stocks of fish being exterminated. This ambiguous conclusion should be compared with the sharp conclusions drawn recently in many articles. (See Beddington *et al.* (1975), Berck (1979), Gould (1979), Peterson and Fisher (1977) and Hoel (1978).) It should be noticed, however, that in these contributions the unrealistic case of a constant number of fishing units is assumed, that is n is assumed to be a constant irrespective of the size and even the sign of profit.

From Figure 11.4 we can see that extermination can only take place if the path for development in n and Z remains in the area (d) right down as far as $Z = 0$. If we assume that the cost of fishing a given quantity tends towards infinity when Z tends towards nought, this cannot occur. The path will be forced into the area (a).

This cost assumption sounds reasonable. However, we cannot exclude the possibility that there are fisheries in which the stocks that yield a zero pure profit (Z^s), or even worse, do not cover all costs, are very small and close to nought. In such cases fishing may continue until stocks are exterminated. An example

287

of fishing of this kind may be stocks of fish occurring in such concentrations that even when stocks are greatly reduced, the fish are easy to catch. If reaction parameter δ in addition is very small, this will reinforce this tendency towards extermination of stocks.

11.6.6 Variations in quantity fished

So far we have made no precise statement with regard to the development in the quantity fished, $n^s x^s$. If we know the development in n and Z, the development of nx can be deduced from (11.13). In area (b) both n and Z will increase. The quantity fished will then also increase. It will continue to increase, too, in area (c). In this area, however, the quantity fished will start to decrease. This reduction will be accelerated in area (d), continuing into area (a). Gradually reduction will come to a halt, and once again the quantity fished will start to increase. This increase will start in area (a), accelerating in area (b). (The reasoning is built on the fact that from (11.13) it follows that when $\dot{n}$ and $\dot{Z}$ both have the same sign, then $(\dot{n}x)$ also has this sign, and that nx is a continuous function of time t.) Table 11.1 below will illustrate this development.

Table 11.1

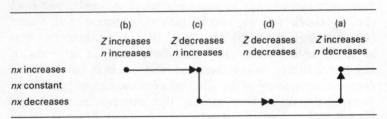

	(b)	(c)	(d)	(a)
	Z increases n increases	Z decreases n increases	Z decreases n decreases	Z increases n decreases
nx increases				
nx constant				
nx decreases				

These variations in fishing are not variations over a short period of time, but more long-term variations. These are not variations that can be confirmed on the basis of statistics for quantities fished, for example, of herring, over a period of the last two to three years, but are based on figures covering a longer period of time, for example, 10 to 20 years. It must be borne in mind that in 'real life' changes in parameters which we have kept constant are always taking place, for example, factor prices, and in parameters which have not been included at all.

A positive shift in product price will result in both the $n(Z)$ curve and Z^s undergoing a negative shift. Point (n^s, Z^s) will move inwards towards origo.

In an extended model we can allow p to be determined by the demand and supply of fish. Supply has been explained above. Demand could be specified by assuming that the price p depends only on the total quantity sold, for example, so that p is a falling convex function of nx. This would change the analysis somewhat, but would not alter the features involved in the kind of unregulated fishing we have discussed so far. In addition, we shall find that over a period of time the price of fish will fluctuate in the opposite direction of variations in the quantity fished. Fluctuations will be more damped than when p was given exogenously.

11.7 EMPIRICAL INDICATIONS: HERRING FISHERIES IN ICELANDIC WATERS

Herring fisheries in Icelandic waters have been based on the exploitation of three stocks of herring (Icelandic summer spawning herring, Icelandic spring spawning herring and mature Norwegian spring spawning herring), also known as 'the Atlanto-Scandian herring'. Together these three species of herring constitute stocks which ensure that the theory set out in the previous sections approximates very closely to these fisheries. From time to time a great many nations have participated in these fisheries — fishermen from Iceland, Norway, the Faeroes (Denmark), Finland, Sweden, the Soviet Union and the German Federal Republic. The large number of fishermen engaged and the large number of nations involved means that the presupposition set out in the previous section for unorganised fishing, with every fisherman a price taker, would appear to be reasonable.

Column 2 in the table below gives the estimate of biologists for development of stocks. Column 3 lists of the total quantity fished by members of the nations mentioned above. Column 4 shown the catch an Icelandic boat has landed as an average. We have had access to data for Icelandic catches and the number of Icelandic vessels. We have, however, not had access to figures showing the number of vessels from other nations participating.

The total number of boats is shown as an estimate in column 5. It is assumed that non-Icelandic vessels on an average achieve the same catch as the Icelandic.

There are certain events in the development of these fisheries which are worth noting: in 1959 a technical development got under way which led to the replacement in 1961 of dories (which up to then had been used in purse seine fishing) by power blocks for drawing together and pulling up the net, while at the same time the actual net is now larger and stronger (nylon having replaced hemp), while practically every boat is also equipped with sonar devices. We also note that the quantity fished from 1961 and for a number of years thereafter is far larger than previously. As from 1964 the fishing of herring in Icelandic waters has been undertaken practically only by Icelanders, a fact that is also connected with the extension to a 12 mile fishing limit. It was not until after 1964 that the full effects of this extension were felt. The total number of boats listed in the fifth column is therefore Icelandic only from 1964. In 1952 the catch per Icelandic boat is surprisingly small, probably due to a short Icelandic season occasioned by labour dispute. The total number of vessels engaged in these fisheries in 1952 is therefore assessed on the basis of Icelandic catches per vessel in 1951. In the first place 1951 approximately in time closely to 1952. For this reason technical factors should differ little from conditions obtaining in 1952. In the second place stocks in 1951 approximate more closely to those in 1952 than, for example, in 1953. On the basis of the presuppositions we made with regard to the cost structure in the preceding section, the structure in 1951 should therefore more closely resemble the structure in 1952 than the cost structure in 1953.

It will be seen from the Table 11.2 that changes in stocks cannot exclusively be explained on the basis of the quantity fished. Changed recruiting and growth conditions caused by environmental changes in the sea help to explain why stocks appear to be vanishing entirely in the course of this 20-year period.

Technical changes and changes in the maritime environment have not been taken into consideration in the theoretical discussion carried out in the preceding section.

If we compare Figures 11.4 and 11.5, we shall see that from 1950 to 1956 we have a development in the actual fisheries

Table 11.2: Herring fisheries in Icelandic waters, 1950-69

Year	Stocks in million tons	Quantity fished 1,000 tons	Catch landed per Icelandic boat, 1,000 tons	Estimated number of boats of the 'Icelandic type'
1950	9.7	74	0.129	573
1951	9.4	106	0.294	361
1952	9.2	61	0.294[a]	208
1953	7.6	95	0.233	407
1954	8.3	61	0.132	462
1955	9.6	75	0.212	354
1956	9.8	124	0.369	336
1957	11.0	143	0.380	376
1958	9.5	151	0.311	486
1959	8.5	238	0.522	456
1960	6.5	224	0.337	665
1961	5.1	462	0.704	656
1962	4.1	650	—[b]	—[b]
1963	2.9	508	0.734	692
1964	2.8	626	1.515	211
1965	3.3	624	2.299	204
1966	2.9	483	3.397	168
1967	1.5	118	1.892	148
1968	0.3	31	0.731	103
1969	0.02	24	—[b]	—[b]

Notes
   a. See text.
   b. Data not available.

Sources
   Jakob Jakobson; 'Exploitation of the Icelandic spring and summer spawning herring in relation to fisheries management 1947-1977', working paper from Marine Research Institute, Reykjavik, 1978; Olav Dragesund, J. Hamre and Ø. Ullfang: 'Biology and population dynamics of the Norwegian spring spawning herring', working paper from Institute of Marine Research, Bergen, 1978; 'Tölfrædihandbók', 1974, Reykjavik, 1976; 'Ægir', Fiskefélag Islands, 1950-69

corresponding to the type of rigmarole indicated in the theoretical fisheries in Figure 11.4. From 1957 on development deviates from the theoretical pattern; stocks decline rapidly while at the same time a large number of vessels are engaged in fishing. Now a deviation is precisely what should have been expected in the 1960s, since, as mentioned above, a well-nigh revolutionary new technique was introduced into the fisheries at the end of the 1950s. The profits from fishing rose, *inter alia*, because costs were depressed. Sonar equipment, too, meant that the increasing costs, which would otherwise have resulted from a marked decline in stock, were very efficiently cushioned. Even though we have not included this formal technical progress in our model above, this explanation to the development set out in Figure 11.4 appears to be reasonable. From 1963 to 1964 it will be seen that the number of boats drop drastically. In reality only Icelanders themselves continued fishing after 1964. As mentioned above this must be seen in connection with the measures introduced in Iceland for regulating fishing, although part of the explanation may also be that stocks had now declined to the level where, despite the use of sonar equipment, transport costs to and from the fishing grounds were sufficiently high to deter all but the Icelanders.

Reports moreover have now come in of stocks which are recovering, though at a very gradual rate. This will undoubtedly mean that before too long fishing for this species of herring will once again be resumed. We must therefore expect that, over a period of time, we shall receive observations of increasing stocks and an increasing number of vessels, presupposing that no revolutionary progress takes place in fishing, similar to the technical changes that were seen at the end of the 1950s.

11.8 REGULATED FISHING

11.8.1 Introduction

In unregulated fishing the problem was that stocks of fish were influenced by the activities of the fishermen. Stocks of fish had a bearing on the fishermen's cost function. Increased fishing, all else being equal, resulted in lower stocks of fish in the future and for that reason smaller catches in the future, and thus higher fishing costs. The individual fisherman has no motive for taking into consideration these negative indirect effects. The quantity fished by each boat constitutes $1/n$ part of the total quantity fished. When n is sufficiently large, the individual will consider it pointless to impose any restrictions on himself, as long as there is no guarantee that others will do the same.

We shall now assume that fishing is organised in such a way that negative external effects will be taken care of. We shall allow a central unit to determine the total quanti'y fished, so that discounted earnings from fishing will be as great as possible. The central authority will take into account the growth

Figure 11.5: Development of stocks of herring and number of boats in the Icelandic herring fisheries

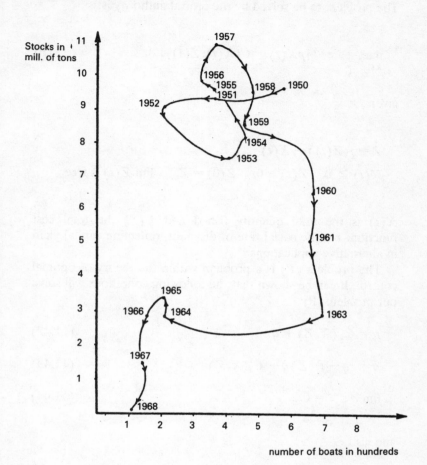

relation (11.6) which shows how stocks of fish are developing. From now on we shall ignore the problem of determining the number of fishing vessels that should be employed.

This simplification enables us to operate with a function for the total quantity. Finally, we shall briefly discuss how the central authority can initiate its fishing policy.

11.8.2 The central planning problem

The problem to be solved by the central authority is:

$$\max_{X(t)} \int_0^\infty e^{-rt} [pX(t) - C(X(t), Z(t))] \, dt \qquad (P)$$

given

$$\dot{Z} = f(Z(t)) - X(t)$$
$$X(t) \geqq 0 \quad Z(t) \geq 0, \quad Z(0) = Z_0, \quad \lim_{t \to \infty} Z(t) \text{ is free}$$

$X(t)$ is the total quantity fished and $C(\cdot)$ the total cost function. r is the social rate of discount, reflecting capital yield in alternative applications.

The problem (P) is a problem within the theory of optimal control. It can be shown that the following conditions will solve our problem (P):

$$p = C'_x(X, Z) + q \qquad (11.42)$$

$$\dot{q} = rq - f'(Z)q + C'_z(X, Z) \qquad (11.43)$$

$$\lim_{t \to \infty} q \, e^{-rt} = 0 \qquad (11.44)$$

and also

$$Z(0) = Z_0$$

The righthand side in (11.42) consists of two components. $C_x(X, Z)$, as before, indicates private marginal costs in fishing, q indicates the value of a unit of uncaught fish. Let

$$\max_{X(\tau)} \int_t^\infty e^{-r(\tau - t)} [pX(\tau) - C(X(\tau), Z(\tau))] \, d\tau = J(Z(t))$$

Then

294

$$q(t) = \frac{\partial J(Z(t))}{\partial Z(t)}$$

that is, equal to the marginal discounted value resulting from stocks of fish having been a little greater at point of time t. The fact that q occurs on the right-hand side of (11.43) reflects the circumstance that there is an alternative use for fish caught, apart from that of ending up on the quay, namely, as an uncaught fish it can swim off and grow bigger for 'next year', and/or ensure that there will be more fish. The sum on the righthand side represents the social marginal costs in fishing. We see that as long as the marginal cost of catching and delivering a fish at the quay is positive, the value of a fish on the quay is greater than that of a fish in the sea. The price of a unit of uncaught fish develops as given in (11.43).

The equation expresses the fact that the gain in rise of price resulting from allowing the marginal fish to remain in the sea is to correspond to the loss of interest resulting from allowing it to remain in the sea minus the value of the change in growth minus the savings in catching costs (via external stock effect).

In a potential stationary state we shall have $\dot{Z} = \dot{q} = 0$. Relation (11.43) will then give us:

$$q^s = - C'_z(X^s, Z^s)/(r - f'(Z^s)) \qquad (11.45)$$

A stationary state with $f'(Z^s) > 0$ is possible. This will imply $Z^s < Z^m$; the optimal regulated stocks are less than the MSY condition. The reason why this could be optimal, even though the amount could have been caught at a lower cost for a $Z > Z^m$, is that we derive utility from the actual reduction in stocks.

With regard to the question of extermination, we can see from (11.42) that this will only be the case if the product price is greater than the sum of the marginal costs of catching the 'last fish' and the shadow price involved in allowing it to remain in the sea.

We shall not continue with the general case, but rather introduce specific functional forms in order to arrive at more explicit results.

11.9 EXPLICIT FUNCTIONAL FORMS

11.9.1 Regulated fishing

We make the following assumption with regard to growth and cost functions:

(i) $\quad f(Z) = Z(\bar{Z} - Z)$

(ii) $\quad C(X, Z) = \dfrac{1}{2}(\dfrac{X}{Z})^2$

(i) has been explained above (see section 11.5); (ii) means that we are discussing only the rising portion of the marginal cost curve and that the passus coefficient in fishing, for a constant stock Z, is very low, namely, 0.5. We then get

$$C'_x = \frac{X}{Z^2}$$

$$f' = \bar{Z} - 2Z$$

This involves discussing the following relations:

$$p = \frac{X}{Z^2} + q \tag{11.42'}$$

$$\dot{q} = [r - (\bar{Z} - 2Z)]\, q - \frac{X^2}{Z^3} \tag{11.43'}$$

$$\dot{Z} = Z(\bar{Z} - Z) - X \tag{11.16'}$$

From (11.42') we know how the instrument variable X at all times will be adjusted according to the size of the state variable Z and the shadow price q:

(a) $\quad X = (p-q)\, Z^2$

(b) $\quad X = 0$ for $p = q$ and for $Z = 0$, since $X \geq 0$ the demand $q \leq p$ emerges

(c) $\quad \dfrac{\partial X}{\partial q} = x'_q = - Z^2 < 0,\; X''_{qq} = 0$ $\qquad\qquad$ (11.46)

(d) $\quad \dfrac{\partial X}{\partial Z} = X'_Z = (p-q)2Z > 0,\; X''_{ZZ} = 2\,(p-q) > 0$

(e) $\quad \dfrac{\partial^2 X}{\partial q \partial Z} = \dfrac{\partial^2 X}{\partial Z \partial q} = X'_{qZ} = - 2Z < 0$

After inserting (11.46) in (11.43′) and (11.16′) we get

$$\dot{q} = [r - (\bar{Z} - 2Z)]\, q - (p-q)^2\, Z \qquad\qquad (11.47)$$

$$\dot{Z} = Z(\bar{Z} - Z) - (p-q)\, Z^2 \qquad\qquad (11.48)$$

These are two differential equations in the two unknown time functions $q(t)$ and $Z(t)$. In adition we know that $Z(0) = Z_0$ and require that (11.44) holds;

$$\lim_{t \to \infty} q(t)\, e^{-rt}\, Z(t) = 0$$

11.9.2 Unregulated fishing as a basis of comparison

In so far as we have a slightly different and more explicit description of fishing in this section than in the previous sections, we shall start the further discussion of the unregulated fishing in this case.

The unregulated fishery emerges by putting

$$q \equiv 0$$

We then get the equations

$$p = \frac{X}{Z^2} \tag{11.42''}$$

$$\dot{Z} = Z(\bar{Z} - Z) - X \tag{11.16''}$$

By substituting for X from $(11.42'')$ we get the differential equation

$$\dot{Z} = Z(\bar{Z} - Z) - pZ^2$$

The development over time of the stock of fish in the case of unregulated fishery can now be described by the phase diagram in Figure 11.6. In the unregulated fishery the stock will increase monotonically towards the stable stationary state Z_u or decrease monotonically towards this stock level. This description of the unregulated fishery corresponds to the description in Figure 11.2 when we also assume a constant number of boats as a simplification. The stock of fish will increase if it is lower than Z_u initially and decrease if initial stocks are greater than Z_u. From this and from $(11.42'')$ we see that when stocks of fish are initially 'big', the unregulated fished quantity will be big to start with and then decline. When stocks of fish initially are small, the quantity caught will be small to start with and then increase. In the stable stationary state the fished quantity will be

$$X_u = pZ_u^2$$

The profit from fishing will in the stationary state be

$$\Pi_u = pX_u - C(X_u, Z_u) = pX_u - \frac{1}{2} \frac{X_u X_u}{Z_u^2} = \frac{1}{2} pX_u$$

11.9.3 Time development for stocks of fish and quantity caught in the case of regulation

Relations (11.47) and (11.48) now apply. We shall first introduce an auxiliary magnitude $\hat{Z}$, which is defined as the Z which ensures that

$f'(\hat{Z})$ is equal to r

Figure 11.6: The stock development of the unregulated fishery
$Z_u = \bar{Z}/(1+p)$

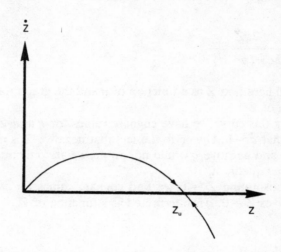

If we multiply by q on both sides we then get

$$qf'(\hat{Z}) = q \cdot r$$

q is the value per unit of fish in the sea. $f'(\hat{Z})$ is the marginal yield of stocks of fish or natural capital. $qf'(\hat{Z})$ is therefore the value of the marginal productivity of natural capital. qr is the user cost involved in utilising the natural capital, provided we ignore the fact that reduced stocks of fish increase the cost of fishing. This cost of utilising fish resources is taken care of by the last component on the righthand side of (11.43).

With the fully specified f-function we have chosen we get

$$\hat{Z} = \frac{\bar{Z} - r}{2}$$

Using the fact that $f'(\hat{Z}) = \bar{Z} - 2\hat{Z}$ and that $Z_u = \bar{Z}/(1 + p)$, then (11.47) and (11.48) can be written as follows:

$$\dot{q} = 2(Z - \hat{Z})q - (p - q)^2 Z \qquad (11.47')$$

$$\dot{Z} = Z[(1+p)(Z_u - Z) + qZ] \qquad (11.48')$$

In order to arrive at the time development for q and Z we need to find the graphs for q as a function of Z when $\dot{q} = 0$ and $\dot{Z} = 0$. By inserting $\dot{q} = 0$ in (11.47') we find that

$$Z = \frac{2q\dot{Z}}{2q - (p - q)^2} \tag{11.49}$$

We shall here find Z as a function of q and the graph for q as a function of Z.

Along this curve we have cognate values for q and Z which ensure that $\dot{q} = 0$. Owing to the fact that negative Z is not permissible and negative q would not occur, we are only interested in the first quadrant.

By substituting $\dot{Z} = 0$ we find cognate values for q and Z which gives $\dot{Z} = 0$. q is interpreted as a function of Z.

$$q = (1 + p)\left(1 - \frac{Z_u}{Z}\right) \tag{11.50}$$

By combining (11.49) and (11.50) we arrive at the phase diagram given in Figure 11.7.

$Z_u = $ stationary state in unregulated fishing $= \bar{Z}/(1 + p)$
$\hat{Z} = $ defined by means of the Z that maximises $q(f(Z) - rZ), f'(\hat{Z}) = r$
$\hat{Z} = (\bar{Z} - r)/2 \approx \bar{Z}/2 = Z^m$. r is a small figure compared with Z
$\bar{Z}$ is the stationary condition for Nature left alone.

$$\bar{q} = 1 + p + \sqrt{1 + 2p}, \underline{q} = 1 + p - \sqrt{1 + 2p}$$

The arrowed curve indicates the fact that optimal time development for q and Z is a development converging on the stationary state (q_{opt}, Z_{opt}). Any other choice would either result in a development in which $Z = 0$ or in the condition

$$\lim_{t \to \infty} q(t) e^{-rt} Z(t) = 0$$

not being fulfilled.

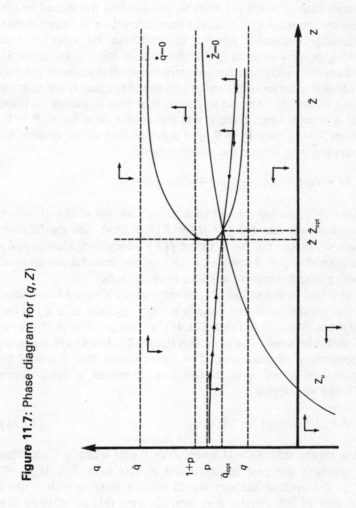

Figure 11.7: Phase diagram for (q, Z)

The stationary state $(q_{opt}, \bar{Z}_{opt})$ is locally asymptotically stable. Global stability has not been investigated.

Characteristics of optimal time development

(1). Assume that Z_0 is very small, but positive. We can then pose the question of whether it would be optimal to impose a ban on fishing. From (11.46a) we can see that this would be the case by choosing $q = p$. X will then, according to decision rule (11.46b), be equal to nought. The condition for a cessation of fishing proving optimal is that the sum of the private profit Π and capital yield in connection with the natural resource fishing, evaluated in terms of the value of a fish in the sea, is less than or equal to nought. An end to fishing that would interest us must last a certain time. During this period we must have $X = 0$. When $X = 0$, then $\Pi = 0$. $q = p$ means that in the considered interval $\dot{q} = 0$, so that the sum mentiond

$$\Pi + (\dot{q}Z + q\dot{Z}) = 0 + 0 + p\dot{Z} = pf(Z)$$

Meanwhile, we have assumed a growth law for stocks of fish of such a nature that $f(Z) > 0$ for $\bar{Z} > Z > 0$. Our model contains no lower limit for Z where $f(Z)$ is negative. Hence, $f(Z)$ for a small $Z_0 > 0$ cannot be zero. In our model a cessation of fishing would therefore not be a relevant policy.

If initial stocks of fish Z_0 are very small, it would be optimal in our model to choose initially a 'high' q_0, that is, a q_0 in the neighbourhood of p. From (11.46) we can see that in that case X would be small. X will be less than $f(Z)$. Stocks are given an opportunity of increasing. The figure shows that it would be optimal to allow X to decline over a period of time. From (11.46) we see that

$$\dot{X} = 2Z(p-q)\dot{Z} - Z^2\dot{q} \qquad (11.51)$$

This means that $\dot{X} > 0$ when $\dot{Z} > 0$ and when $\dot{q} < 0$. This is precisely the case for an initial Z_0 less than $\bar{Z}_{opt}$. If $Z_0 < Z_{opt}$, the optimal strategy would involve starting with a small quantity of fish caught, and then allowing this quantity of fish caught to increase in the direction of the level $\bar{X}_{opt} = (p - \bar{q}_{opt})$ $\bar{Z}_{opt}^2$. Corresponding to this choice of optimal time development for the quantity fished we have a 'penalty', q, for fishing that does not exist in the case involving unregulated fishing. In the area under consideration this q starts at a high level, gradually declining in the direction of level $\bar{q}_{opt}$. In this area stocks of fish increase monotonically in the direction of level $\bar{Z}_{opt}$.

(2) We assume that the initial stock Z^0 is so great that $Z^0 > \bar{Z}_{opt}$. The optimal strategy would then be to choose a 'low' q_o, and then allow q to increase over a period of time. Optimal stocks of fish are initially 'big', decreasing over time towards the level $\bar{Z}_{opt}$. The corresponding optimal quantity of fish caught is to start with great, but then declines monotonically in the direction of $\bar{X}_{opt}$.

(3) It follows that the optimal $q(t)$ will at all times be less than p, but greater than $q = (1+p) - \sqrt{1+2p}$. It is perhaps more interesting to note that the optimal stationary stocks of fish $\bar{Z}_{opt}$ must *necessarily* be greater than $\hat{Z}$. $\hat{Z}$ is equal to the Z that for given q and r maximises

$$qf(Z) - qrZ$$

This $\hat{Z}$ in our model is virtually speaking equal to (actually slightly less than) the Z maximising $f(Z)$.

In the figure we have chosen p in such a way that $Z_u < \hat{Z}$. This may not necessarily be the case. If

$$p < \frac{\hat{Z}+r}{\hat{Z}-r}$$

then $Z_u > \hat{Z}$.

What, however, is bound to happen is that $\bar{Z}_{opt} > Z_u$. This follows from the figure. The optimal stationary stock, $\bar{Z}_{opt}$, must in other words of necessity be greater than the stationary stocks in unregulated fishing, Z_u, and stock Z^m that ensure maximal natural yield.

11.9.4 Ensuring optimal fishing

There are various ways in which optimal fishing can be ensured:

(A) Fishing vessels may be allocated fishing quotas.
(B) A charge or duty may be levied on each fishing vessel, based on the quantity fished.

303

By investigating the adjustment in unregulated fishing, it will be seen that if each vessel has to pay a charge or duty equal to q per unit of the quantity fished, then each vessel will arrange things in such a way that

$$p = c_i'/Z + q$$

This is precisely the condition characterising optimal fishing.

From the above it will be seen that this fishing duty or levy can either start at a low rate (large initial stocks of fish) and then increase in the direction of the stationary level $\bar{q}_{opt}$, or start at a high level, and then decline (small initial stocks of fish) towards the stationary level $\bar{q}_{opt}$.

REFERENCES AND FURTHER READING

Beddington, J.R., C.M.K., Watts and W.D.C. Wright (1975): 'Optimal cropping of self-reproducible natural resources', *Econometrica*, 43, 789-802

Berck, P. (1979): 'Open access and extinction', *Econometrica*, 47, 877-82

Dragesund, O., J. Hamre and Ø. Ullfang (1975): 'Biology and population dynamics of the Norwegian spawn herring', Working paper from the Institute of Marine Research, Bergen

Gould, J.R. (1979): 'Extinction of fishery by commercial exploitation: a note', *Journal of Political Economy*, 80, 1031-8

Hanneson, R. (1978): *Economics of fisheries*, Universitetsforlaget, Oslo

Hoel, M. (1978): 'Extermination of self-reproducible natural resources under competitive conditions', *Econometrica*, 46, 219-24

Jakobson, J. (1978): 'Exploitation of the Icelandic spring and summer spawning herring in relation to fisheries management 1974-1977', working paper from Marine Research Institute, Reykjavik

Munro, G.R. and A.D. Scott (1985): 'The economics of fisheries management', in *Handbook of Natural Resources and Energy Economics*, (eds.) A.V. Kneese and J.L. Sweeny, vol. 2, North Holland, Amsterdam

Neher, P.A. (1974): 'Notes on the Volterra-quadratic fishery', *Journal of Economic Theory*, 8, 39-49

Peterson, F.M. and A.C. Fisher (1977): 'The exploitation of extractive resources. A survey', *The Economic Journal*, 87, 681-721

Smith, V.L. (1969): 'On models of commercial fishing', *Journal of Political Economy*, 77, 187-98

Tølfrædihandbók 1974, Reykjavik, 1976

Ægir, Fiskefélag Islands, 1950-69

Name Index

Subject Index